ORGANIC POLYMER
CHEMISTRY: A PRIMER

A Supplement to
ORGANIC CHEMISTRY
by William H. Brown

BRUCE M. NOVAK

Department of Polymer Science and Engineering
University of Massachusetts

Saunders Golden Sunburst Series

SAUNDERS COLLEGE PUBLISHING
Harcourt Brace College Publishers

Fort Worth   Philadelphia   San Diego   New York
Orlando   Austin   San Antonio   Toronto
Montreal   London   Sydney   Tokyo

Novak: Organic Polymer Chemistry: A Primer: A Supplement to
Organic Chemistry by William H. Brown
ISBN 0-03-010633-8

56 095 98765432

Preface

Organic Polymer Chemistry: A Primer is intended for use as a supplement to Brown, *Organic Chemistry*. The purpose of this chapter is to give students an introduction to the important and ever-expanding field of polymer chemistry. While some material on polymers is presented in Chapters 5 and 20 of *Organic Chemistry*, this chapter contains a more in-depth treatment of molecular weights, molecular weight distributions, polymer morphology, step-growth and chain-growth polymerizations, ring-opening metathesis polymerizations, and conjugated polymers through ROMP techniques.

Like the textbook, this chapter contains worked-out examples followed by similar problems for students to try on their own. It also has a summary of all key reactions and additional end-of-chapter problems. To illustrate the applications of polymer chemistry, three *Chemistry in Action* essays are presented: *Polymers in Medicine; The Development of Ziegler-Natta Catalysts*; and, *The Recycling of Plastics*.

I would like to take this opportunity to thank a number of people for helping me prepare this manuscript: Lisa Boffa for writing the *Chemistry in Action* essays; Professor Roger Porter, my UMass colleague, for many helpful discussions; and, Professors James Mulvaney of the University of Arizona, and William H. Brown of Beloit College for reading the entire manuscript and providing insightful comments. Finally, I extend a very special thanks to my lovely wife, Julie Novak, for once again lending her superb editorial skills to my manuscripts.

CONTENTS

Organic Polymer Chemistry: A Primer

The technological advancement of any society is inextricably tied to the materials available. Indeed, historians have used the emergence of new materials as a means of establishing a timeline in order to mark the development of human civilization. In terms of current technologies, many potentially desirable breakthroughs await the availability of novel raw materials with specific properties. As part of a general push aimed at identifying such materials, scientists have made increasingly greater use of organic chemistry in the preparation of synthetic polymers. This represents an exciting period for organic chemists as they apply their knowledge of reactions and reaction mechanisms to the development of polymers with very specific properties. The unusual versatility afforded by organic chemistry allows for the creation and elaboration of materials at a level previously unattainable using traditional wood, metals, or ceramics. Deceptively simple changes in the chemical structure of a given polymer can change its mechanical properties from those of a typical sandwich bag to something more reminiscent of a bullet-proof vest. In fact, we will see later that even very simple organic polymers can be far stronger and stiffer than even steel. Further structural changes can introduce properties never before imagined in organic polymers. For example, using well-defined organic reactions, polymers can be converted from insulators (e.g., the rubber sheath that surrounds electrical cords) to electrical conductors with conductivities nearly equal to that of copper metal!

Polymers are long chain molecules synthesized by linking monomers together through chemical reactions. The molecular weights of polymers can be quite high in comparison to organic molecules commonly encountered in the laboratory and typically range from 10,000 g/mol to more than 1,000,000 g/mol. The architectures or shapes of macromolecules can also be quite diverse, and these in turn have important influences on their properties. Examples of different shapes include linear and branched polymers. In addition to these common architectures, specialty materials with star, comb, and ladder structures (Figure 1) can also be synthesized, and these polymers exhibit their own unique properties.

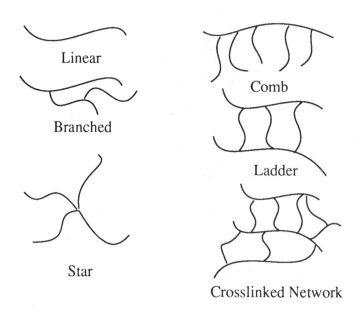

Figure 1: Various polymer architectures.

Further structural variations can be achieved by introducing covalent bonds, called crosslinks, between the individual polymer chains. Crosslinked polymers form networks that have very different properties from those of linear or branched structures. Linear and branched polymers are often soluble in typical solvents such as chloroform, benzene, toluene, DMSO, or THF. In addition, many linear and branched polymers can be melted into highly viscous liquids. Polymers of this type that can be melted are called **thermoplastic**, which means that they will flow when heated. In contrast, crosslinked polymers are insoluble and infuseable materials that cannot be melt processed. Crosslinked polymers are sometimes called **thermosets** because their shapes are set by the crosslinks that prevent the polymer chains from flowing upon heating. Because of these very contrary physical characteristics, thermoplastics and thermosets must be processed differently, and are used in very different applications.

1. POLYMER NOTATION AND NOMENCLATURE

By convention, polymers are depicted by placing parentheses around the **repeat unit** or **mer,** which is the smallest molecular fragment or grouping that contains all of the nonredundant structural features of the chain. It is understood by this notation that the structure of the entire chain can be reproduced by repeating the enclosed structure in both directions. A subscripted variable called the **average degree of polymerization** is placed outside the parentheses to indicate that this unit is repeated n number of times.

One notable exception to this structural notation is in the case of polymers formed from symmetric monomer units such as polyethylene, $(-CH_2CH_2-)_n$, or poly(tetrafluoroethylene), $(-CF_2CF_2-)_n$ (Teflon). Although the simplest repeat units are the $-CH_2-$ and $-CF_2-$ groups respectively, convention dictates showing both methylene groups (or both difluoromethylene groups) originating from the ethylene ($CH_2=CH_2$) or tetrafluoroethylene ($CF_2=CF_2$) monomer.

As is the case for organic chemistry in general, the nomenclature of polymers is not fully systematic and the actual nomenclature in use is a mixture of common and IUPAC names. The three methods of naming polymers are (1) nomenclature based on the monomer, (2) trade names and acronyms, and (3) nomenclature based on the IUPAC system. The prevalence of trade names and acronyms in this field is symptomatic of its origins in industrial laboratories. The most common method of naming a polymer is to attach the prefix *poly* to the name of the monomer. For example, polyethylene and polystyrene.

3

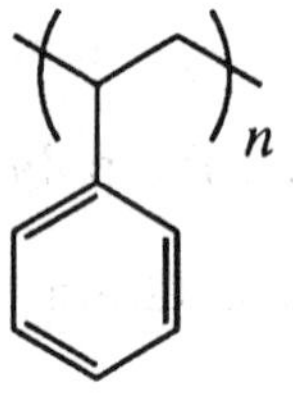

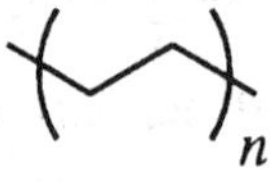

Polystyrene   Polyethylene

In the case of a more complex monomer, parentheses are used, as in poly(vinyl alcohol) or poly(vinyl chloride). Whenever possible it is preferable to use systematic IUPAC rules to name the monomer.

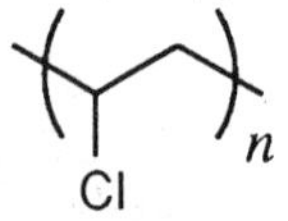

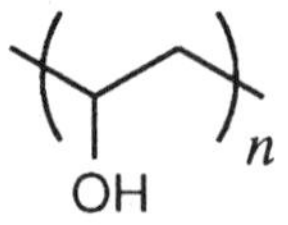

Common Name:	PVC	Vinylon or PVA
Name based on Monomer:	Poly(vinyl chloride)	Poly(vinyl alcohol)
IUPAC Name:	Poly(chloroethylene)	Poly(hydroxyethylene)

To name polymers systematically using the IUPAC system, the simplest repeat unit in the polymer, sometimes called the **constitutional repeat unit (CRU)**, is identified. In the case of symmetric monomers, the CRU is not the same as the monomer repeat. For example, poly(tetrafluoroethylene) or Teflon, has the monomer repeat structure $(-CF_2CF_2-)_n$, but the simplest CRU is $-CF_2-$. Therefore, the IUPAC name based on this smallest subunit is poly(difluoromethylene).

Example 1

Given the following structure; (a) determine the polymer's repeat unit and redraw the structure using the simplified parenthetical notation, and (b) name the polymer using IUPAC nomenclature rules.

Solution

(a)   The polymer has regularly alternating $-CH_2-$ and $-CF_2-$ groups and, therefore, the simplest repeat unit must contain both of these groups:

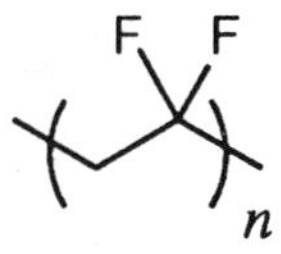

(b)   Using IUPAC guidelines, the repeat unit is derived from 1,1-difluoroethylene, and therefore, the polymer would be called poly(1,1-difluoroethylene).   Also called poly(vinylidene fluoride), this polymer is used in microphone diaphragms.

Example 2

Name the following polymer using the IUPAC system.

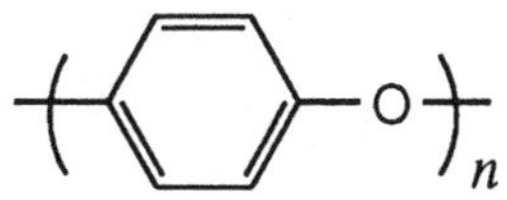

Solution

This polymer, with the oxygen and phenyl group within the repeat structure is called poly(oxy-1,4-phenylene). Its common name is poly(phenylene oxide).

Problem 1

From the structure shown below, identify the repeat unit of the polymer and name it according to IUPAC rules.

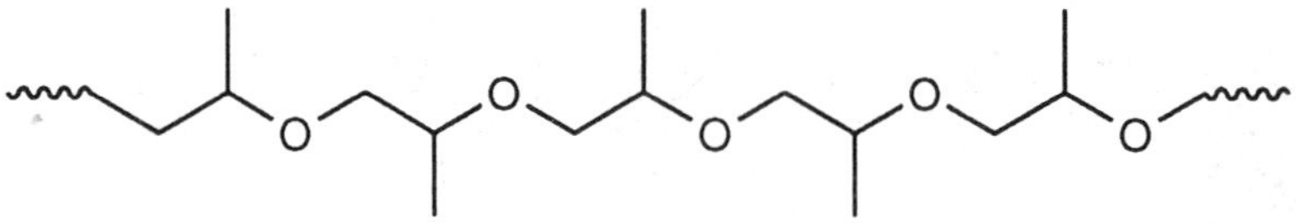

2. AVERAGE MOLECULAR WEIGHS AND MOLECULAR WEIGHT DISTRIBUTIONS

The single most important attribute of macromolecules is their size, and as a direct result of this unique characteristic, macromolecules may be endowed with properties such as elasticity, strength, and adhesion that are not commonly associated with small organic molecules. While it is true that intramolecular covalent bonding and weak intermolecular forces are exactly the same for large and small molecules, large molecules have many more intermolecular contacts per molecule, so that the sum of the forces holding the molecules together is appreciable. Furthermore, long-chain polymer molecules tend to become entangled with one another in much the same way as cooked spaghetti behaves in your kitchen strainer. These multiple intermolecular contacts and chain entanglements make it difficult for individual polymer molecules to move past one another and deformations become more difficult. A good example of the importance of size is a comparison of paraffin wax and polyethylene. These two highly distinct materials actually have identical CRUs (i.e., $(-CH_2-)_n$) but differ greatly in chain size. Paraffin wax is composed of chains having 25 to 50 carbon atoms, whereas the polyethylene chains average between 1000 and 3000 carbons per chain. Paraffin wax such as in birthday candles tends to be weak and brittle, but polyethylene, as in a plastic milk bottle, is strong, flexible, and tough. These vastly different properties arise directly from the difference in size of the molecular chains.

All synthetic polymers and some naturally occurring polymers are actually composed of a mixture of individual polymer molecules of variable molecular weights. The breadth of this molecular weight distribution in a given sample is important not only in determining the polymer's properties, but also because it contains information about the synthetic process used in preparing the sample. Given a sample of any set of objects, there are several ways in which to measure an average value for the distribution. When defining molecular weights in polymer chemistry, the two most common averages used are the **number average (M_n)** and **weight average (M_w) molecular weights**. The number average, M_n, given by equation 1, is the common arithmetic mean calculated by counting the number of chains of a particular size, summing these, and then dividing this sum by the total number of chains (just as your average grade in a series of organic examinations is calculated by adding all grades and dividing by the total number of exams).

$$M_n = \frac{\sum n_i M_i}{\sum n_i} \qquad (1)$$

n_i = number of chains with molecular weight M_i
$\sum n_i$ = total number of chains

The weight average molecular weight (M_w) is calculated by recording the total weight of each chain of a particular length, summing these weights, and dividing by the total weight of the sample (equation 2).

$$M_w = \frac{\sum w_i M_i}{\sum w_i} = \frac{\sum n_i M_i^2}{\sum n_i M_i} \qquad (2)$$

$w_i = n_i M_i$ = weight of chains with molecular weight M_i

The use of these two different methods for calculating the average molecular weight of a sample stems from the fact that there are experimental techniques that essentially count particles (colligative properties, for example) that yield M_n and other techniques that measure the size of particles (light scattering) that yield M_w. Because the larger chains in a sample weigh more than the smaller chains, the weight average is skewed to higher values and M_w is always greater than M_n (Figure 2).

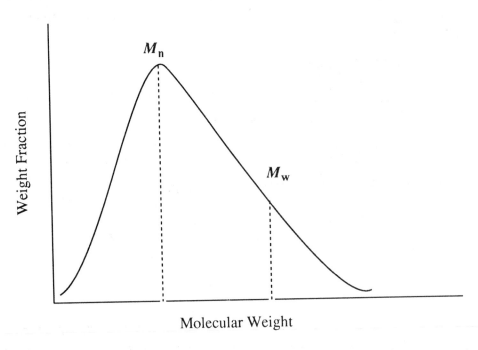

Figure 2: The distribution of molecular weights in a given polymer sample.

Example 3

Given the following data, calculate the number average molecular weight (M_n) of this polymer sample.

Fraction	Mole Fraction	Molecular Weight
1	0.05	17,000
2	0.07	24,000
3	0.09	38,000
4	0.12	52,000
5	0.20	61,000
6	0.11	77,000
7	0.10	85,000
8	0.08	90,000
9	0.05	100,000
10	0.05	110,000
11	0.04	121,000
12	0.04	134,000

Solution

In equation 1, the term $n_i/\Sigma n_i$ is equal to the mole fraction of chains at molecular weight M_i. Therefore:

M_n = 0.05(17,000) + 0.07(24,000) + 0.09(38,000) + 0.12(52,000) + 0.20(61,000) +

0.11(77,000) + 0.10(85,000) + 0.08(90,000) + 0.05(100,000) + 0.05(110,000) +

0.04(121,000) + 0.04(134,000)

M_n = 69,000

Problem 2

Using the molecular weight data in Example 3, calculate the M_w for this sample.

Both M_n and M_w are intrinsically useful values, and their ratio, M_w/M_n, sometimes called the **polydispersity index** provides a measure of the breadth of the molecular weight distribution. As mentioned, the size distribution of polymer chains in a given sample is important in determining its properties: In the solid-state, low-molecular-weight chains tend to plasticize, or soften, a polymer, making it less able to resist deforming forces. Higher-molecular-weight chains, because they have the ability to entangle with one another to a greater extent than smaller chains, greatly increase the viscosity of material's liquid melt. This increased viscosity influences the processing of thermoplastic materials during molding applications.

When the M_w/M_n ratio is unity, all of the polymer molecules are the same length and the polymer is said to be **monodisperse.** No man-made polymers are ever monodisperse unless the individual molecules are carefully fractionated using time-consuming, rigorous separation techniques based on molecular size. On the other hand, natural polymers, such as polypeptides and DNA, that are formed using biological processes are monodisperse polymers. This is one of the attributes of these biological polymers that lends them their unusual and highly specialized properties.

3. POLYMER MORPHOLOGY: CRYSTALLINE VERSUS AMORPHOUS MATERIALS

Driven by thermodynamics, polymers, like small organic molecules, tend to crystallize upon precipitation or as they are cooled from the melt. Acting to inhibit this tendency are their high molecular weights, which tend to slow diffusion, and their sometimes complicated or irregular structures which prevent efficient packing of the chains. As the net effect of this play-off, polymers in the solid-state tend to be composed of both ordered, or crystalline, domains and disordered, or amorphous, domains. Amorphous domains are characterized by the absence of long-range order. Highly amorphous polymers are sometimes referred to as **glassy** polymers. The relative amounts of crystalline and amorphous domains differ not only from polymer to polymer but also depend upon the manner in which the material was processed. High degrees of crystallinity are more often found in polymers with regular, compact structures, and strong intermolecular forces such as hydrogen bonding and dipolar interactions.

The degree of crystallinity within a polymer contributes greatly to its properties and, as such, figures prominently in determining its applications. Amorphous polymers are transparent materials because they lack crystallites that scatter light. In terms of mechanical strength, they are typically weaker polymers, both in terms of their flexibility and their strength. As the degree of crystallinity increases in a sample, the polymers become increasingly more opaque due to scattering of the light by the crystallites. Associated with the introduction of crystalline domains, is usually a corresponding increase in their strength and stiffness. This increased performance is due to the crystallities acting as crosslinks between the individual polymer molecules. This relationship between mechanical properties and the degree of crystallinity can be illustrated by poly(ethylene terephthalate) (PET).

Poly(ethylene terephthalate) (PET)

PET can be made with percent crystallinities ranging from zero to about 55%. Completely amorphous PET is formed by quickly cooling from the melt. By slowing down the cooling time, more molecular diffusion can occur and crystallites begin to form as the chains begin to order. The differences in mechanical properties between these two forms of PET are great. PET with a low degree of crystallinity is used for plastic soda bottles, whereas highly crystalline PET fibers can be used for strong fabrics and tire cords.

Crystalline domains in polymers undergo melt transitions, abbreviated T_m, in a fashion exactly analogous to the melting of simpler organic compounds. Amorphous phases, however, do not undergo this transition; instead, they are transformed from a hard glass to a rubbery state. The temperature of this latter transition is called the **glass transition temperature,** or T_g. The distinction between these transitions can best be viewed by considering the changes that occur upon cooling a polymer from a high-temperature melt. As a liquid polymer begins to cool, molecular energy in the form of translations, vibrations, and rotations decreases. When these molecular motions have diminished significantly, intermolecular forces between chains become more important and if the chains have a sufficiently regular structure, they will begin to order and crystallization occurs. Looking at this same phase transition from the opposite direction, the temperature at which crystallites melt corresponds to the T_m of the polymer. For example, poly(6-aminohexanoic acid) $(-CONH(CH_2)_5-)_n$, a polymer with uses ranging from textile fibers to shoe heels, has a $T_m = 223°C$. At room temperature (and well above), this is a hard durable material that does not undergo any appreciable change in properties even on a very hot summer afternoon, which makes it ideal for these applications.

As energy is removed by cooling from a melted polymer that lacks structural regularity, a point eventually will be reached at which there is no longer sufficient energy for main chain motions of the polymer to continue and the polymer will enter a glassy state. The temperature at which main chain motions cease is the glass transition temperature. Amorphous polymers at temperatures below their glass transition temperatures are hard, glassy solids, and above this temperature they are, soft, flexible rubbers. Amorphous polystyrene with a $T_g = 100°C$ is a rigid solid at room temperature (the rigid plastic used for drinking cups, foam packaging materials, disposable medical wares, tape reels, etc.), but if placed in boiling water, it becomes soft and rubbery. Natural rubber, polyisoprene, $(-CH_2C(CH_3)=CHCH_2-)_n$, has a $T_g = -73°C$ and is of course a rubbery solid at room temperature.

Completely amorphous polymers exhibit only a glass transition temperature. Most polymers are composed of both amorphous and crystalline domains, and, therefore, show both a T_m and a T_g transition. For example, PET has a $T_m = 270°C$, and a $T_g = 69°C$. The temperature values of these transitions are also important in determining the properties of the polymers and in establishing the range of temperatures over which the polymer will be useful. Values of T_g and T_m for polymers tend to vary considerably. In the absence or minimization of restrictions to rotation around the polymer's main chain, as in the case of polyethylene, $(-CH_2CH_2-)_n$, T_g values can be very low, (e.g., $-125°C$ for polyethylene). Polydimethylsiloxane, $(-Si(CH_3)_2O-)_n$ sometimes called silicone, with its relatively long Si—O bond (0.164 nm vs. 0.153 nm for a carbon-carbon single bond), has both a low T_g and T_m ($-127°C$ and $-40°C$, respectively). At

12

room temperature, which is above both its glass transition and melt temperatures, silicone is a viscous fluid. If the polymer is pulled or otherwise distorted, the individual polymer chains elongate and slip past one another, and the sample stretches and eventually pulls apart, much like a piece of taffy. If, however, crosslinks are introduced between the individual chains, the material behaves as an elastomer. When a low T_g, crosslinked polymer is pulled, the chains also begin to elongate, but now the crosslinks prevent the chains from pulling apart. When the distorting force is released, extended chains tend to return to their original conformations. This elastic recovery is driven by entropy. One of nature's fundamental driving forces is to increase the disorder, or entropy, in a system. Coiled polymer chains have more entropy (less order) than extended chains, because fully extended chains have only one optimum conformation associated with the rotational bond angles along the main chain. In polyethylene, the fully extended chain has the staggered, zig-zag conformation (See Section 2.5A of *Organic Chemistry*.).

Trans, zig-zag conformation
(low entropy)

Random coil
(high entropy)

In contrast, coiled polymer chains are highly disordered, having random rotational angles along the main chain. Hence, driven by entropy, flexible polymer chains naturally tend to adopt the more plentiful random-coil conformations. When polymer chains are stretched, their entropy decreases, and upon release, the chains fold back into the more probable random coil conformations, thereby increasing their entropy. For this reason, elastomers are sometimes referred to as *entropy springs*.

Rubber materials must have low T_g values in order to behave as elastomers. If the temperature drops below this T_g value, then the material is converted to a rigid glassy solid and all elastomeric properties are lost. A less than full appreciation of this idiosyncratic behavior of polymers was a contributing factor in the Challenger spacecraft disaster in 1985. The elastomeric O-rings used to seal the solid booster rockets had relatively high T_g values (around 0°C). When the temperature dropped to an unanticipated low on the morning of the Challenger launch, the O-ring seals dropped below their T_g value and obediently changed from elastomers to rigid glasses, losing any sealing capabilities. The rest was tragic history.

4. STEP-GROWTH POLYMERIZATIONS

Polymerizations in which chain growth occurs in a slow, stepwise manner are called **step-growth polymerizations,** or **condensation polymerizations** (Section 20.12). The "step" terminology is used because it accurately reflects the kinetics of the process wherein high-molecular-weight polymers are assembled in a stepwise fashion as monomers react to form dimers. In turn dimers form tetramers, tetramers become octomers, and so on. This orderly growth scheme is naturally complicated by fragments of any and all sizes reacting with one another (monomer with tetramer, for example). Many of the synthetic polymers on the commercial market are formed through step-growth polymerizations. These materials range from thermoplastic polyamides, polyesters, and polycarbonates used as fibers and cloth to common thermosets such as epoxy- and melamine-formaldehyde resins (Formica countertops). Step-polymers are formed by allowing difunctional monomers with complementary functional groups to react with one another. There are two common types of step-growth processes, involving either the reaction between A-A and B-B type monomers or the self-condensation of A-B monomers. In both cases the A functional groups react exclusively with B groups and B groups exclusively with A groups. An A-A/B-B type step-growth process can be illustrated by the reaction of ethylene glycol and terephthalic acid to form the common polyester poly(ethylene terephthalate) (PET), used in the manufacture of recyclable plastic soda bottles and textile fabrics (equation 3).

Terephthalic acid   Ethylene glycol   Poly(ethylene terephthalate)

The self condensation of 6-aminocaproic acid provides a good illustration of an A-B type step-growth process (equation 4).

6-Aminohexanoic acid
(ε-aminocaproic acid)
          Poly(6-aminohexanoic acid)

The resulting polymer, poly(6-aminohexanoic acid), also called *nylon-6*, is a common polymer used as fibers, brush bristles, rope, high-impact moldings, tire cords and in other similar applications.

This stepwise method of construction of polymer chains has important consequences for both the molecular weights and the molecular weight distributions of the polymers produced. Probability tells us that the most abundant species tend to co-condense; thus, at the reaction's inception, small chains most likely react with other small chains. This tendency persists to very high conversion, and consequently, high-molecular-weight polymers are not produced until very late in the reaction (i.e., past 99% conversion) when there is finally a probability of larger chains reacting with one another. This statistically-determined molecular weight profile means that only very high yielding reactions can be used to form step-growth polymers. This restriction points to an important distinction between small molecule organic reactions and step-growth polymerizations. Although a reaction that typically yields 85% of the desired product is considered "good" in organic synthesis, the same reaction is essentially useless for step-growth polymerizations, because high-molecular-weight polymers will never be formed at such low conversions. Furthermore, monomer purity takes on added importance because any impurities, such as monofunctional molecules, added to the chain ends effectively deactivate the chain toward further growth. This endcapping of polymer chains by monofunctional molecules is sometimes used synthetically to purposely limit the molecular weight of polymers for specific applications in which large molecules would be inappropriate.

Step-growth polymerizations can be subdivided into two additional classes: *condensation* and *step-addition*. Condensation reactions, illustrated previously by the formation of PET and nylon-6 (equations 3 and 4), eliminate byproducts during the bond-forming steps. These byproducts might include water, alcohol, halohydric acids, or alkali metal halides depending upon the structure of the starting monomers. In step-addition polymerizations, the bond forming steps proceed without generation of byproducts. Examples of step-addition polymerizations include the formation of polyureas from the reaction between diisocyanates and diamines (equation 5), and the formation of epoxy resins from the reaction of diepoxides and diamines (equation 6).

$$OCN\!-\!\boxed{}\!-\!NCO \;+\; H_2N\!\diagdown\!\diagup\!NH_2 \longrightarrow \left[\!\!\begin{array}{c} \end{array}\!\!\right]_n \quad (5)$$

1,4-Diisocyanatobenzene   1,2-Diaminoethane   Poly(ethylene phenyl urea)

$$\text{A diepoxide} \quad + H_2N\!\diagdown\!\diagup\!NH_2 \longrightarrow \qquad\qquad (6)$$

A diepoxide   A diamine

An epoxide resin

4.A. Mechanisms for Step-Growth Polymerizations

The mechanism for the formation of a typical polyester from the reaction between a diol and a diacid is illustrated in equations 7 through 9 and involves the attack of the alcohol nucleophile on the electrophilic carbon of the carboxyl group to form a tetrahedral intermediate that decomposes into the ester and water (see Fischer esterification, Section 19.9B).

16

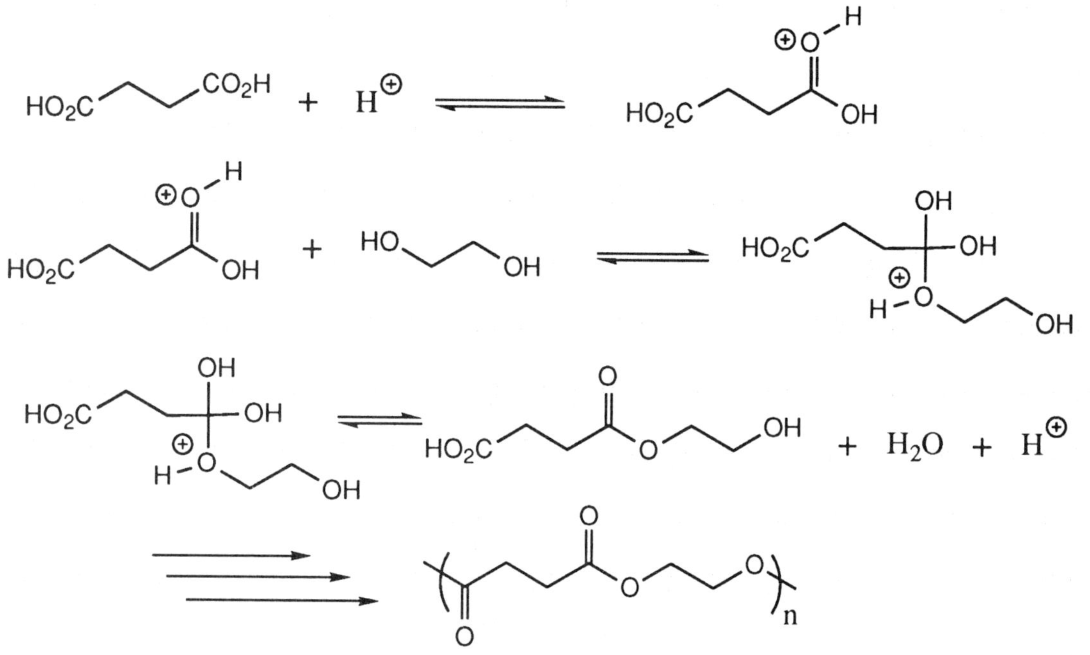

(7)

(8)

(9)

These reactions are reversible and typically require the removal of water in order to drive the reaction in the forward direction.

Example 4

Write a detailed mechanism for the acid-catalyzed polymerization of 1,4-diisocyanatobenzene and ethylene glycol.

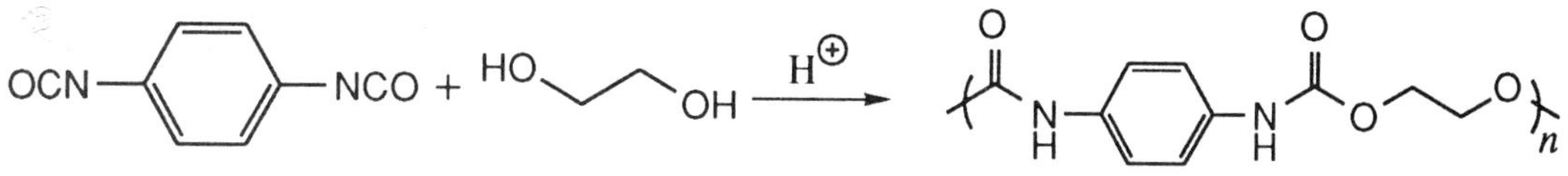

1,4-diisocyanatobenzene   Ethylene glycol   A polyurethane

Solution

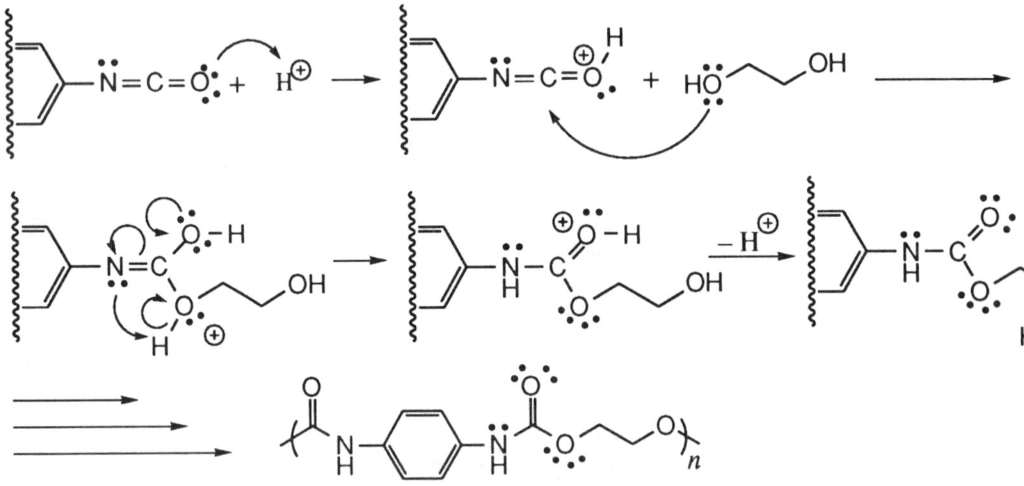

Problem 3

Predict the product from the following reaction and write a detailed mechanism showing its formation.

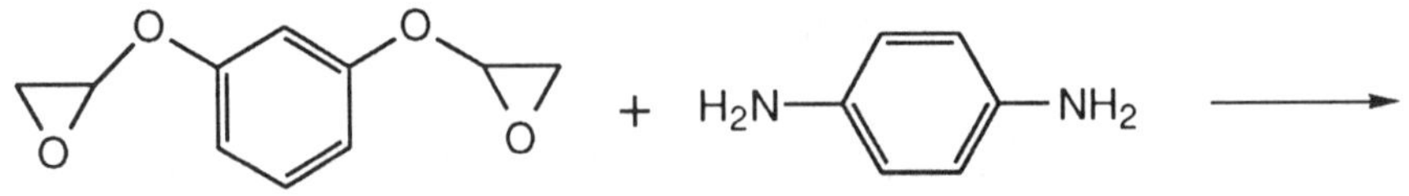

A diepoxide   A diamine

The mechanism of polymerization in step-growth processes is identical to the reactions of small molecules except for the obvious difference that ever-larger chains are reacting with one another as the reaction proceeds. An important question to ask is whether, because of diffusional effects, the rate of reaction between functional groups slows down as the chains become larger. Interestingly, for most cases the answer is no. This conclusion is based in part on kinetic results obtained from the study of the esterification reaction of a series of homologous aliphatic carboxylic acids differing from each other only in chain length (equation 10).

$$H\text{-}(CH_2)_x\text{-}CO_2H \ + \ CH_3CH_2OH \ \longrightarrow \ H\text{-}(CH_2)_x\text{-}CO_2CH_2CH_3 \ + \ H_2O \qquad (10)$$

$x = 0, 1, 2,...$

These studies show that the esterification rates initially decrease with increasing size, but become essentially constant for molecules larger than butanoic acid ($x = 3$). This result lends support to the idea that the polymer chain-ends continue to react at good rates even very late in the step-growth process when large changes in the molecular weight occur and the solution becomes more viscous.

4.B. Applications of Step-Growth Polymers

Some of the first synthetic polymers were condensation polymers, prepared by allowing diacids to condense with diamines to form amide bonds and water, and with diols to form ester bonds and water. Polyamides or nylons, formed from the condensation of diamines and diacids, still hold an important place in today's synthetic materials markets. Their properties are diverse and depend upon the chemical structure of the repeat units. For example, poly(6-aminohexanoic acid) (equation 4) exhibits desirable properties which make it useful in a number of demanding applications such as rope or tire cords. These properties can be improved, however, by replacing the aliphatic segments with aromatic segments. This effect is nicely illustrated in the formation of the polymer known as *Nomex* from 1,3-diaminobenzene and 1,3-benzenedicarboxylic acid chloride, as shown in equation 11.

1,3-diamino-   1,3-benzene-
   benzene   dicarboxylic acid chloride   (An aromatic polyamide or aramid)

Nomex crystallizes more efficiently than the flexible aliphatic chains of nylon-6 due to favorable stacking interactions between the aromatic groups. Additionally, the strong carbon-carbon bonds of the aromatic groups coupled with the polymer's highly crystalline nature, lend good thermal stability to this polymer. These properties in combination with the hydrogen bonding of the amide groups, work in concert to yield a material with high strength and heat resistant properties. This polymer is suitable for high strength, high temperature applications such as parachute cords and jet aircraft tires. The *meta*-linkages in Nomex introduce "kinks" in the chains that prevent them from extending out and crystallizing efficiently. Low crystallinity is desirable from a processing standpoint because the polymers are more soluble in polar organic solvents, but it is detrimental to the polymer's overall mechanical properties. The introduction of *para*-linkages in the monomers results in polymers that are more difficult to process but that have highly improved properties. The polymer formed from 1,4-diaminobenzene and 1,4-benzenedicarboxylic acid chloride is called *Kevlar*, and it is used in a number of high performance applications, including cloth for bullet-proof vests (equation 12).

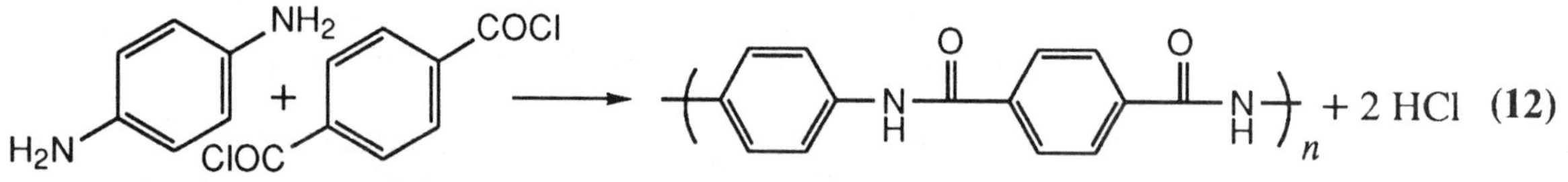

1,4-diamino-   1,4-benzene-
benzene   dicarboxylic
                     acid chloride

Kevlar

Several other very high yielding organic reactions can be used to form step-growth polymers. One important example is the formation of polyimides from dianhydrides and diamines (equation 13).

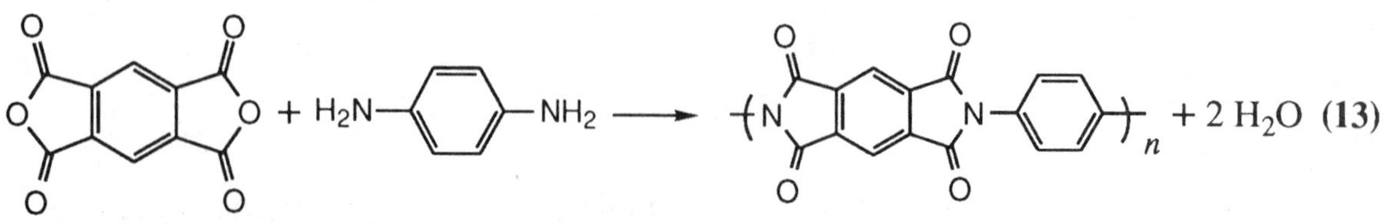

A dianhydride   1,4-Diaminobenzene

Kapton
A polyimide

20

The polymer called *Kapton* shown in equation 13 figures prominently in the electronics industry where it is used as an insulating layer in circuit boards due to its low dielectric constant and its very low tendency to absorb moisture.

Polycarbonates can be formed in two ways, either by allowing phosgene to condense with diols (equation 14), or by using a nucleophilic aromatic substitution route (Section 16.3B) involving aromatic difluoro monomers and carbonate anion (equation 15).

A diphenol   Phosgene   Lexan
A polycarbonate

$$\text{A diphenol} + \text{Phosgene} \longrightarrow \text{Lexan (A polycarbonate)} + 2\,HCl \quad (14)$$

An aromatic
difluoride   Sodium
carbonate   A polycarbonate

$$\text{An aromatic difluoride} + Na_2CO_3 \longrightarrow \text{A polycarbonate} + 2\,HF \quad (15)$$

Poly(4,4'-carbonato-2,2'-diphenylpropane) (*Lexan*) (shown in equation 14) is a tough transparent polymer with high impact and tensile strengths, and is commonly used in the manufacture of safety glass and bullet-proof windows.

Chemistry in Action

Polymers in Medicine

Medical science has advanced very rapidly in the last few decades. Some procedures considered routine today, such as organ transplantation and the use of lasers in surgery, were unimaginable 60 years ago. As the technological capabilities of medicine have grown, the demand for synthetic materials that can be used inside the body has increased as well. Polymers have many of the characteristics of an ideal biomaterial: they are lightweight and strong, inert or biodegradable depending on their chemical structure, and have physical properties (softness, rigidity, elasticity) that are easily tailored to match those of natural tissues. Carbon-carbon backbone and other degradation resistant polymers are used widely as permanent organ and tissue replacements, whereas, those with easily broken linkages are found in degradable stitches and other short-term devices.

Ideally, damaged organs and tissues can be replaced with healthy ones from a donor or from another part of the body. However, the low availability and frequent immunological rejection of donated organs have made synthetic replacements more successful than natural ones in many cases. Heart valves constructed of silicone rubber (polydimethylsiloxane) balls in steel cages or circular flaps of polyoxymethylene are given to more than 60,000 patients annually.

$$\left(\hspace{-4pt} \sim\!\!O \hspace{-4pt} \right)_{n}$$

Polyoxymethylene

Porous Teflon (Gore-Tex) or Dacron polyester fabrics have been used for many years to repair weak or blocked blood vessels. One of the greatest achievements of modern medicine, the artificial heart, relies on the special properties of polyurethanes. These incredibly resilient elastomers, used in the pumping diaphragm and linings of the device, must flex 42 million times each year without degradation or mechanical failure to simulate the function of a healthy heart.

Many other parts of the human body can be reconstructed with polymeric materials. Faulty hip joints are repaired by creating a new polyethylene ball for the ball-and-socket arrangement and attaching it to the femur with poly(methyl methacrylate) cement. A hard, transparent plastic, poly(methyl methacrylate) is used for intraocular contact lenses. The material most commonly used as a replacement for soft and cartilaginous tissues is silicone. Changing the degree of polymerization, the identity of the organic side chains, and the amount of crosslinking present, allows silicone to be prepared in any form from a liquid to a hard rubber. This versatility allows it to substitute for such diverse structures as tendons, finger joints, nose and chin cartilage, and breast tissue.

Although silicone traditionally has been considered fully compatible with the human body, recent findings concerning the low-molecular weight liquid silicone used in breast implants have brought this belief into question. The liquid was found to bleed through the outer casing of the implants and to collect in certain areas of the body, possibly causing connective tissue and autoimmune diseases. The Food and Drug Administration

4B. Applications of Step-Growth Polymerizations

reviewed the safety of silicone implants in 1992 and decided to allow their continued use on a controlled basis. The connection between liquid silicone and these diseases, as well as the implications of these findings for other uses of silicone in the body, may not be apparent for several years.

Although most medical uses of polymeric materials require biostability, applications have been developed that utilize the degradable nature of some macromolecules. Condensation polymers, which are typically made up of ester or amide linkages, are susceptible to aqueous or enzymatic hydrolysis *in vivo*. In most cases the molecules formed during their decomposition are harmless and can simply be metabolized or excreted. The use of poly(glycolic acid) and glycolic acid/lactic acid copolymers as absorbable sutures is an example.

Poly(glycolic acid)	A poly(glycolic acid) - poly(lactic acid) copolymer

Traditional materials such as catgut must be removed by a health care specialist after they have served their purpose, but stitches consisting of these polyacids are hydrolyzed slowly over a period of approximately two weeks. The torn tissues have healed by the time the stitches have degraded, and no removal is necessary.

Both inert and biodegradable polymers have been used in the fabrication of controlled-release drug devices, which consist of a solid polymer matrix containing a pharmaceutical. After being implanted in the body, the drug either diffuses out of the matrix at a controlled rate or is released slowly as the polymer decomposes. These systems offer many advantages over injections or oral administration, including a constant rate of drug release over extended periods and ease of removal when medication is no longer needed. One commercially available controlled-release device is the Ocusert tablet. When implanted in the eyelid, this ethylene/vinyl acetate copolymer capsule continually releases the antiglaucoma drug pilocarpine. Another example is the Norplant contraceptive device, which became available only recently. It consists of six small tubes of silicone rubber that are implanted in a woman's upper arm. The tubes contain a reservoir of the hormone levonorgestrel, which slowly diffuses through the rubber. The six tubes can supply enough levonorgestrel to prevent pregnancy for five years.

5. CHAIN-GROWTH POLYMERIZATIONS

Chain-growth polymerizations involve sequential addition reactions, either to unsaturated monomers or to monomers possessing other reactive functional groups. Both conceptually and phenomenologically, this approach differs greatly from step-growth processes. In the latter, all monomers plus the polymer chain ends possess equally reactive functional groups, allowing for all possible combination reactions to occur (monomer with chain end, monomer with monomer, and chain end with chain end, for example). In contrast, chain-growth processes involve active chain ends possessing intermediates that react with monomer. The monomers themselves, possess functional groups but they do not react with one another. This limits the reactions in these systems to those between the active chain ends and the monomers, exclusively, and alternative pathways, such as the reaction between two monomers, are eliminated. Side reactions between two reactive chain ends do sometimes complicate matters and are discussed later. As a consequence of simplifying the reaction manifold, chain-growth polymerizations are capable of producing high-molecular-weight polymers relatively early in the reaction, long before all of the monomer is consumed.

The reactive intermediates used in chain-growth polymerizations include radicals, carbanions, carbocations, and organometallic complexes. As the name implies, these are chain processes; therefore, when a reactive intermediate reacts during a propagation step, a new reactive intermediate is generated with the result that there is no net change in the concentration of reactive centers. A typical reaction sequence for the free-radical polymerization of a vinyl monomer is shown in Scheme I.

Scheme I

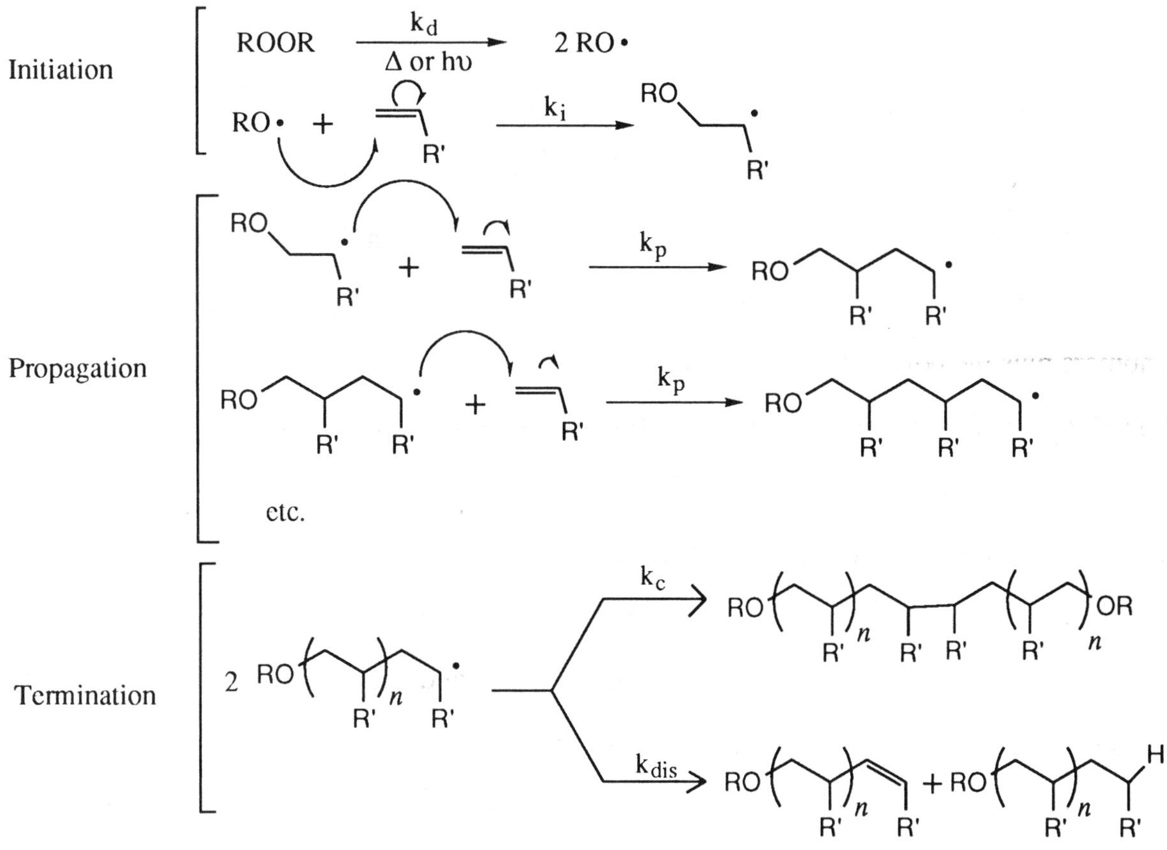

Initiation

Propagation

etc.

Termination

The reaction steps in a chain growth process are divided into three stages: initiation, propagation, and termination. The first two steps are classified as initiation steps and involve transformation of the initiator into a radical essentially identical to the chain carrying radicals present in the propagating steps.

Chain-growth polymerization is a common and important way of producing polymers both in the laboratory and on the industrial scale. The number of monomers that undergo chain-growth polymerizations is large and includes such compounds as alkenes, alkynes, allenes, isocyanates, isocyanides, and strained ring compounds such as lactones, lactams, ethers, and epoxides (Figure 3).

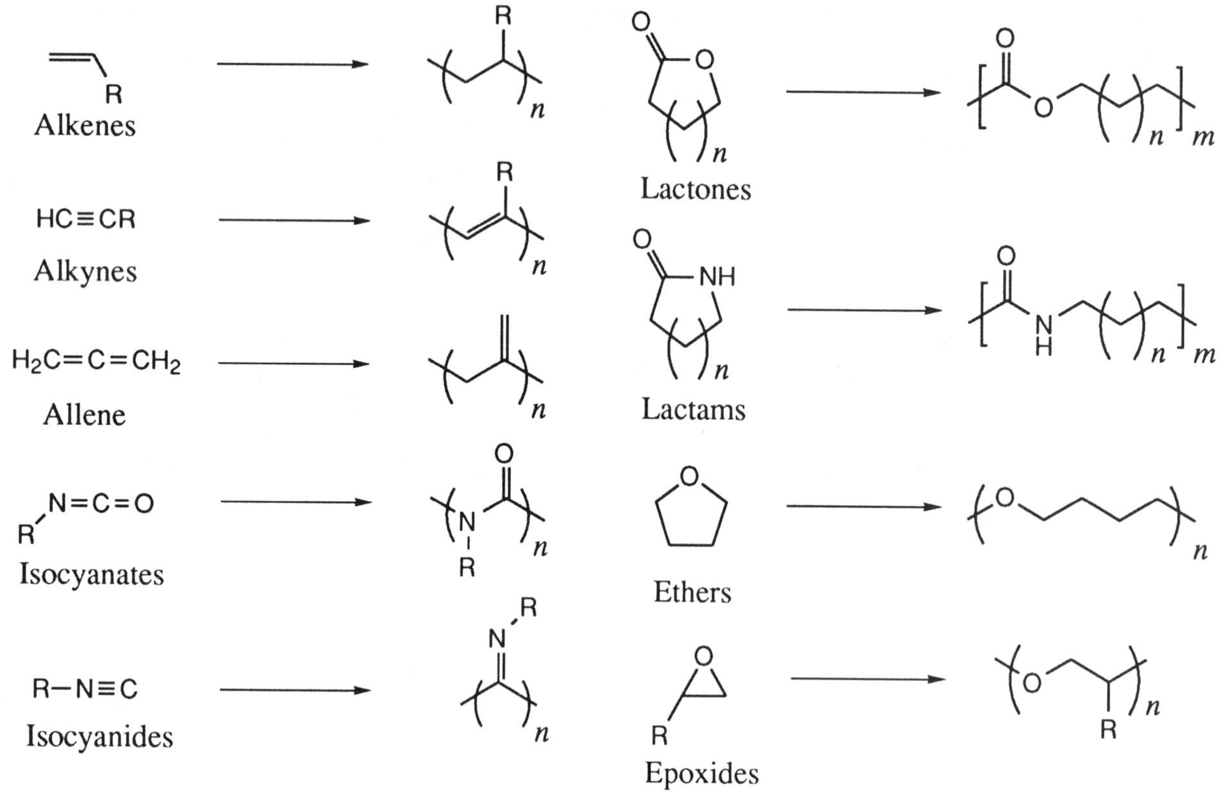

Figure 3: Typical chain-growth polymerization monomers and their polymers.

5.A. Stereochemistry and Polymers.

As in small-molecule organic chemistry, stereochemistry also can play an important role in determining the properties of polymers. Important stereochemical issues are raised during the chain-growth polymerization of substituted alkenes. During the addition step, a reactive sp^2 hybridized center can react through either of its faces with an alkene to form a new stereocenter (Scheme II).

Scheme II

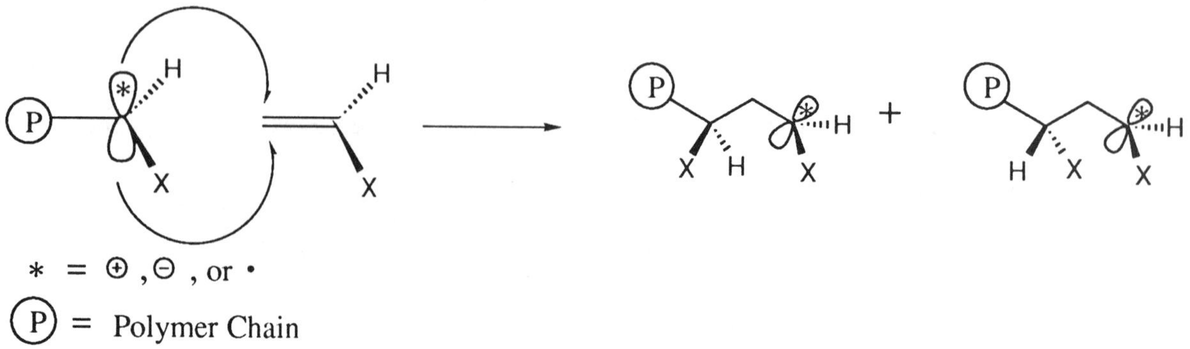

* = $\oplus$, $\ominus$, or •

(P) = Polymer Chain

In the absence of additional stereocenters, these two addition modes are equal energetically, and a racemic mixture will result. Like small molecules, polymers with stereocenters can be chiral molecules. This can be illustrated for short chains using Fisher Projections.

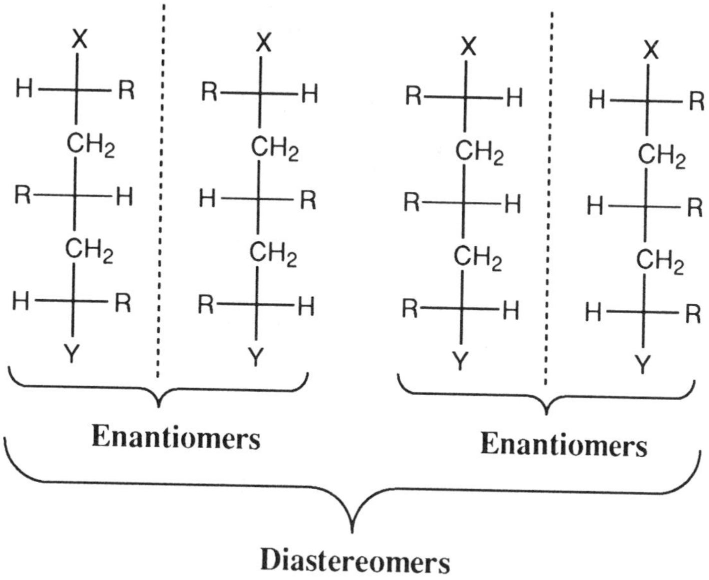

Enantiomers   **Enantiomers**

Diastereomers

A question that arises is whether, if a stereochemical bias is introduced into the system to favor one mode of addition over the other (i.e., favoring one enantiomer over the other), can optically active polymers be formed? The answer in most cases is no. Although stereocenters are formed during the addition reaction, the high-molecular-weight polymers that result from this chain process have a "pseudo-mirror plane" that bisects each of the chiral centers. This mirror plan is called pseudo because it would only be a true mirror plane if the two endgroups of the chain, X and Y, were identical. This pseudo-mirror plane is introduced because of the large size of the polymer chains that extend from either side of the chiral center. As the chain becomes infinite in length, these two "groups" effectively become equal to one another.

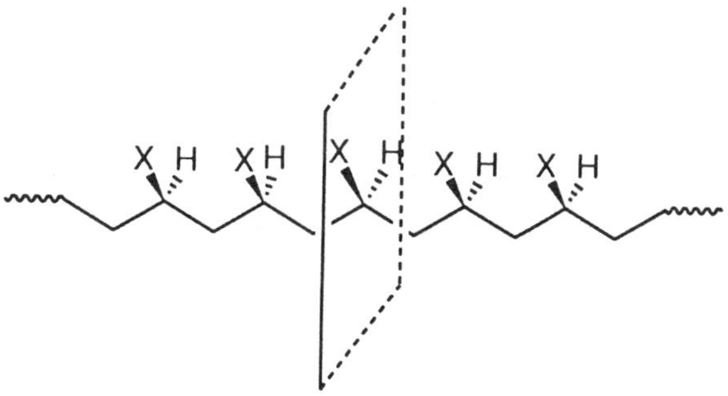

Pseudo-mirror plane

27

Because these pseudo-symmetry elements are present, simple vinyl polymers cannot be optically active due solely to the stereocenters in the backbone. Nevertheless, the relative configurations of stereocenters along a chain are important in determining the properties of a polymer. Polymers with identical configurations at all stereocenters along the chain are called **isotactic** polymers; those with alternating configurations are called **syndiotactic** polymers; and those with completely random configurations are called **atactic** polymers (Figure 4).

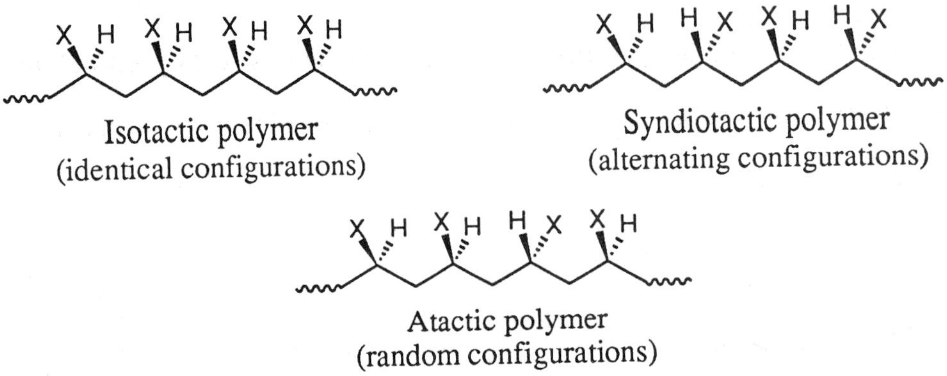

Isotactic polymer
(identical configurations)

Syndiotactic polymer
(alternating configurations)

Atactic polymer
(random configurations)

Figure 4: Relative configurations of stereocenters in polymers with different tacticities.

In general, the more stereoregular the centers are (e.g., in highly isotactic or highly syndiotactic materials), the more crystalline the polymer will be. A random placement of these substituents, such as in atactic materials, results in a polymer that cannot pack well and is usually highly amorphous. This can be illustrated by comparing the properties of polystyrene of differing tacticities. Atactic polystyrene is an amorphous glass with a $T_g \approx 80 - 90\ °C$, whereas isotactic polystyrene is a crystalline fiber-forming polymer with a high melt temperature. The control over the relative stereochemistry, or tacticity, along a polymer backbone is, therefore, an area of important interest in modern polymer synthesis.

Optically-active polymers can be formed by the polymerization of optically-active monomers as long as the stereocenters are not racemized during the polymerization. An example of this would be the polymerization of optically pure (R)-propylene oxide.

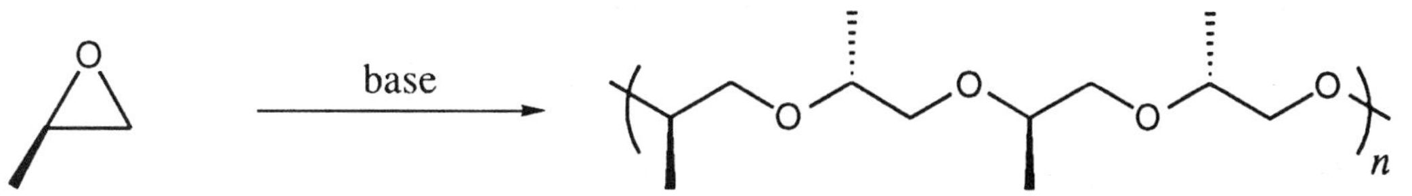

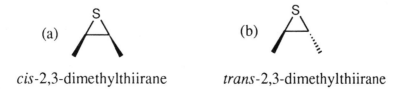

(R)-Propylene oxide   Poly((R)-propylene oxide)

Example 5

Draw the structure of the polymer resulting from the base-catalyzed polymerization of (S)-(+)-lactide, the cyclic dimer of (S)-(+)-lactic acid. Would you expect this polymer to be optically active?

Solution

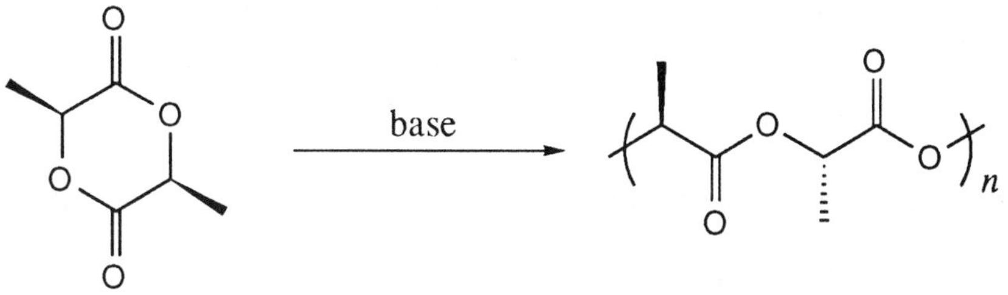

(S)-(+)-Lactic acid   Poly((S)-(+)-lactic acid)

As long as the base is not strong enough the effect racemization of the monomer by deprotonation at the alpha carbon, the stereocenters of the monomer are intact and the resulting polymer will be optically active.

Problem 4

Draw the polymers resulting from the polymerization of the two thiirane monomers shown below. Would you expect either of these polymers to be optically active?

(a)

cis-2,3-dimethylthiirane

(b)

trans-2,3-dimethylthiirane

The properties of polymers with repeat units that can adopt different isomeric forms, or configurations, will also change depending on these forms. For example, polybutadiene can be synthesized in four different pure isomeric forms: 1,4-*cis*, 1,4-*trans*, 1,2-isotactic, and 1,2-syndiotactic.

Isomeric forms of polybutadiene

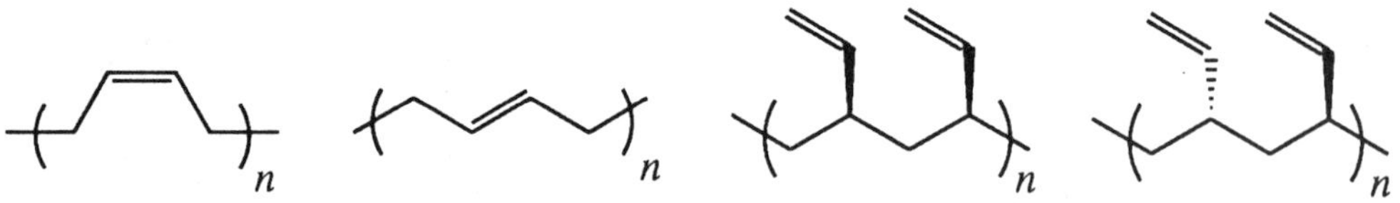

	1,4-*cis*	1,4-*trans*	1,2-isotactic	1,2-syndiotactic
T_g	$-102\,°C$	$-10\,°C$	$-65\,°C$	—
T_m	$6\,°C$	$148\,°C$	$125\,°C$	$154\,°C$

Unless extraordinary care is taken during its preparation, most polybutadiene has varying amounts of these different forms. These polymers are best described as predominantly 1,4-*cis*, or predominantly 1,4-*trans*, and so on. From their different T_g and T_m values, it would be expected that these different isomeric polymers are quite different from one another. In addition, their percent crystallinity is also quite different. Chains containing the 1,4-*cis* units, pack very poorly and 1,4-*cis*-polybutadiene is nearly completely amorphous. Coupled with its low T_g and T_m values, 1,4-*cis*-polybutadiene is a viscous, transparent liquid at room temperature. 1,4-*trans*-Polybutadiene is highly crystalline, indicating that the 1,4-*trans* double bonds pack much more efficiently. As a result, 1,4-*trans*-polybutadiene is an opaque solid at room temperature. Likewise, polymers with either of the regular configurations of the 1,2-polybutadiene are also crystalline solids.

Example 6

Poly(2-hydroxybutanoic acid), a biodegradable polyester, is an insoluble, opaque polymer that is difficult to process into shapes. In contrast, the copolymer between 3-hydroxybutanoic acid and 3-hydroxyoctanoic acid is a transparent polymer that shows good solubility in a number of organic solvents. Discuss the differences between these two polyesters in terms of their structures.

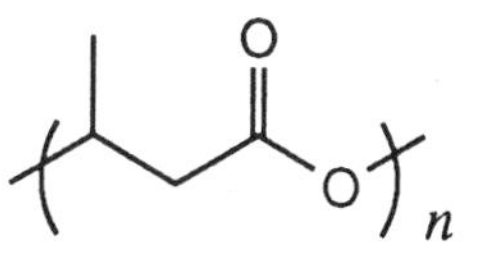

Poly(3-hydroxybutanoic acid) 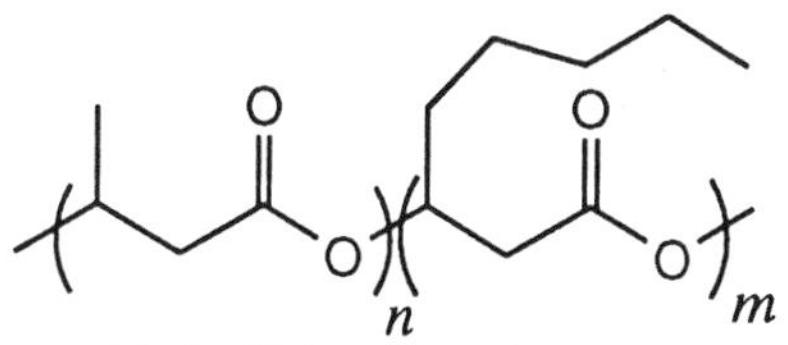   Poly(3-hydroxybutanoic acid - 3-hydroxyoctanoic acid) copolymer

Solution

Any structural irregularity that is introduced into a polymer chain will tend to suppress crystallinity. The lower the crystallinity, the more transparent and soluble the polymer. In this example, the longer C_6H_{13} group acts to disrupt the packing between the polyester chains. This reduces the crystalline content of the polymer making it more processable.

Problem 5

Poly((S)-(+)-lactic acid) is a useful polymer because of its degradability. However, during melt processing steps, it decomposes near its melting point at 180°C. Suggest possible ways of solving this problem.

5.B. Free-Radical Chain-Growth Polymerizations.

Many of the commercial chain-growth polymers are synthesized using free-radical techniques. The free-radical initiators are usually either peroxides or diazo compounds with weak covalent bonds capable of undergoing homolytic cleavage.

Acyl peroxides such as dibenzoyl peroxide typically decompose upon heating in a two-step process. In the first step, homolytic cleavage of the weak O–O peroxide bond yields two acyloxy radicals. These acyloxy radicals then decompose to form two aryl (or alkyl) radicals and CO_2 (equation 16).

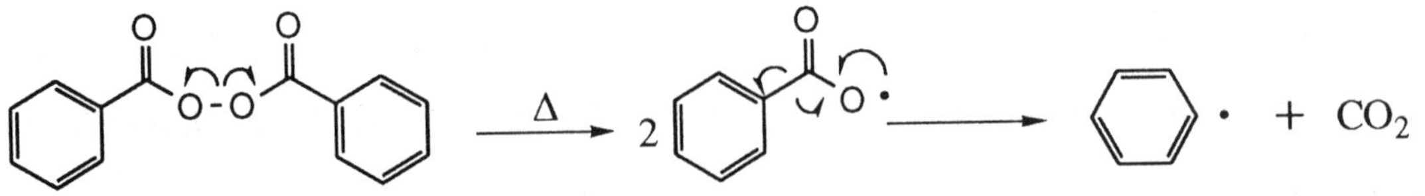

Dibenzoyl peroxide

Another common class of initiators used in radical polymerizations are diazo compounds such as AIBN, which decompose upon heating or by the absorption of UV light to produce two organic radicals and nitrogen gas (equation 17).

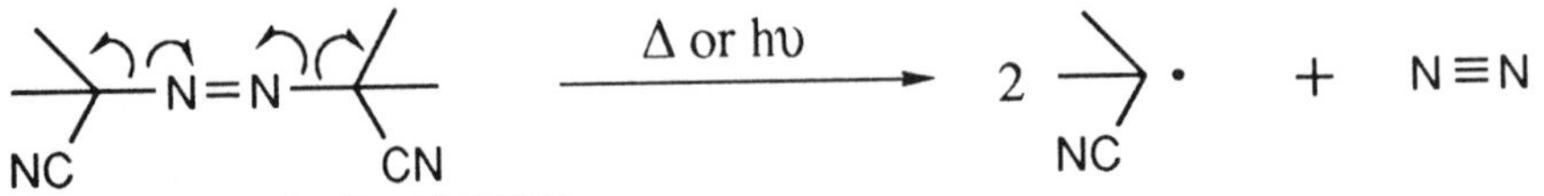

Azodiisobutyronitrile (AIBN)

The organic radicals thus generated then act to initiate polymerization by addition to the double bond of the monomer. Once initiated, the chains continue to propagate through successive additions of the chain-carrying radical to additional monomer. Radical additions to double bonds occur in such a way as to always give the more stable (more substituted) radical. Because additions are biased in this fashion, the polymerizations of vinyl monomers tend to yield polymers with head-to-tail microstructures.

Head-to-tail linkages   Head-to-head and tail-to-tail linkages

Most vinyl polymers made from free-radical processes have no more than 1 to 2% head-to-head linkages. For example, the percentage of head-to-head linkages in poly(vinyl acetate) has been determined to be 1.1% by an acid-catalyzed transesterification (Section 20.5C) to yield poly(vinyl alcohol) and then subjecting this polyhydroxylated material to periodate oxidation (Section 9.5H). Recall that periodate oxidation cleaves only 1,2-diols so the polymer is cleaved at every head-to-head linkage but all of 1,3-diol head-to-tail linkages remain untouched. By monitoring the molecular weight before and after this oxidative cleavage, the number of head-to-head linkages can be determined.

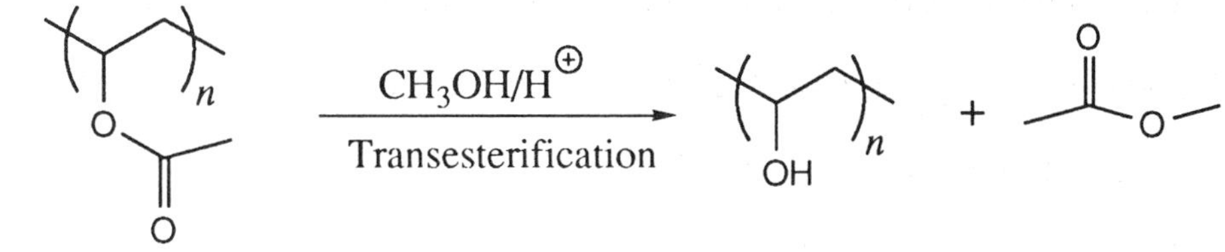

Poly(vinyl acetate)   Poly(vinyl alcohol) Methyl acetate

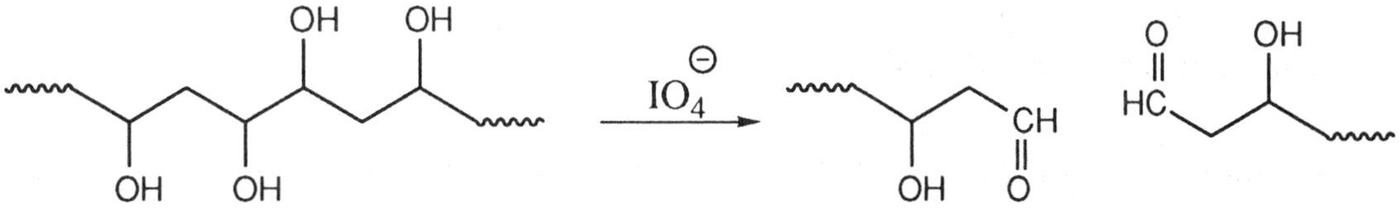

A 1,2-diol, head-to-head   Polymer oxidatively cleaved
linkage in poly(vinyl alcohol)   at 1,2-diol linkage

The chains continue to grow until termination occurs. Termination steps are distinguished from chain initiation and chain propagation steps in the overall scheme by the fact that the active chain ends react in such a way as to consume the reactive endgroups. In radical reactions, the termination process is bimolecular and involves reaction between two radical chain ends. One possibility is simple combination of two radicals to form a new carbon-carbon bond

33

linking the two chains. The combination of two radicals is a facile, diffusion-controlled process that occurs without an activation barrier. In order to suppress this unwanted reaction, the concentration of active radical endgroups is kept low ($\approx 10^{-9}$ to 10^{-7} M). Another common bimolecular termination process, **disproportionation,** involves the abstraction of a hydrogen atom in the beta position to the propagating radical of one chain by the radical endgroup of another chain. This process results in two dead chains, one terminated in an alkane and the other in an alkene (see Termination steps in Scheme I).

Because free-radicals are highly reactive species, it is not too surprising that free-radical polymerizations are very often complicated by unwanted side reactions. A frequently observed side reaction is *hydrogen abstraction* by the radical endgroup from another polymer chain, a solvent molecule, or another monomer. These particular abstraction reactions are not termination reactions, for although they do terminate one chain, they also generate a new radical that is capable of initiating the growth of a new chain. Thus, there is no net change in radical concentration. These side reactions are termed **chain-transfer reactions** because the activity of the endgroup is "transferred" from one chain to another. The ramifications of these reactions are many and include the formation of many chains per single initiator molecular and the introduction of branching into the polymer chains. Transfer to monomer is illustrated for poly(vinyl acetate) in equations 18 and 19.

OAc = CH_3CO_2— Vinyl acetate

"Macromonomer"

34

The newly generated unsaturated radical can then initiate the growth of new chains possessing alkene endgroups. These "macromonomers" can also be polymerized, leading to branched polymer chains (equation 20).

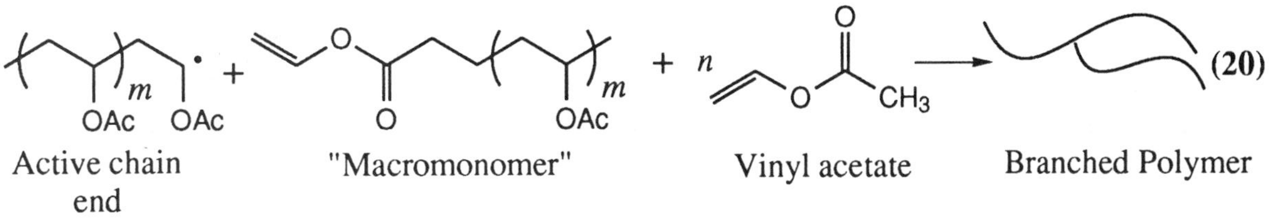

| Active chain end | "Macromonomer" | Vinyl acetate | Branched Polymer | (20) |

Hydrogen atom abstraction can also occur along the polymer's main chain. In such a case the new active radical can begin to grow a new chain, leading once again to branched architectures (equation 21).

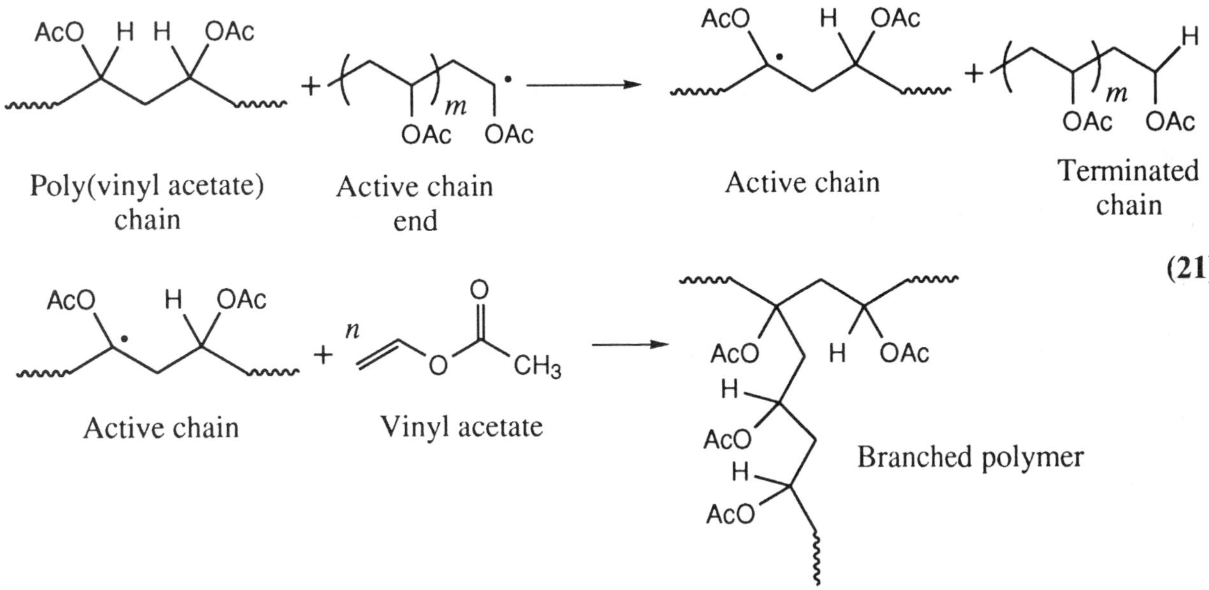

Poly(vinyl acetate) chain   Active chain end   Active chain   Terminated chain

(21)

Active chain   Vinyl acetate   Branched polymer

Polyethylene formed by means of a free-radical process exhibits a number of butyl branches on the polymer main chain. These four-carbon branches are generated in a "back-biting" reaction in which the radical endgroup abstracts a hydrogen from the fifth carbon back (a 1,5-hydrogen abstraction). Continued polymerization of monomer from this new radical center leads to branches four carbons long. Abstraction of hydrogen from the fifth carbon is particularly facile because the transition state associated with this process can adopt a six-member conformation (equation 22).

Six-member transition state leading
to 1,5-hydrogen abstraction

$H_2C=CH_2$ (22)

As a result of these various abstraction reactions, polymers synthesized by means of free-radical processes can have highly branched structures. The number of butyl branches depends on the relative stability of the propagating-radical endgroup and varies depending on the polymer. Polyethylene chains propagate through highly reactive primary radicals which tend to be highly susceptible to 1,5-hydrogen abstraction reactions, and these polymers typically have 15 to 30 branches per 500 monomers. In contrast, polystyrene chains propagate through substituted benzyl radicals that are stabilized by delocalization over the phenyl ring. These stabilized radicals are less likely to undergo hydrogen abstraction reactions and polystyrene typically exhibits only one branch per 4,000 to 10,000 monomers. The presence (or absence) of branches along polymer main chains has important ramifications for the final properties of the polymers. This point can be illustrated by comparing the properties of polyethylene synthesized by free-radical processes with polyethylene obtained from "Ziegler-Natta polymerizations".

5.C.   Ziegler-Natta Chain-Growth Polymerizations.

Karl Ziegler and Guilio Natta shared the 1963 Nobel Prize in chemistry for their discovery of transition metal catalysts capable of initiating the polymerization of ethylene and propylene. The early Ziegler-Natta catalysts were highly active, heterogeneous catalysts composed of Group IVB transition metal halides, a $MgCl_2$ support, and alkylaluminum compounds (i.e., $TiCl_4/Al(CH_2CH_3)_2Cl$), which act to polymerize ethylene at 1 to 4 atmospheres of pressure and at room temperature (equation 23).

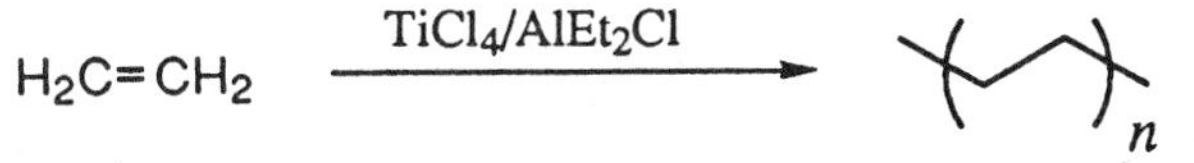

$$H_2C=CH_2 \xrightarrow{\text{TiCl}_4/\text{AlEt}_2\text{Cl}} \left[\text{\Large$\diagup\!\!\!\diagdown\!\!\!\diagup\!\!\!\diagdown$} \right]_n \qquad (23)$$

These heterogeneous mixtures are quite complex, but the active species in a Ziegler-Natta polymerization is thought to be an alkyltitanium compound that is formed by alkylation of the titanium halide by $Al(CH_2CH_3)_2Cl$ on the surface of a $MgCl_2/TiCl_4$ particle. Once formed, this species will repeatedly insert ethylene into the titanium ethyl bond to yield polyethylene (Scheme III).

Scheme III

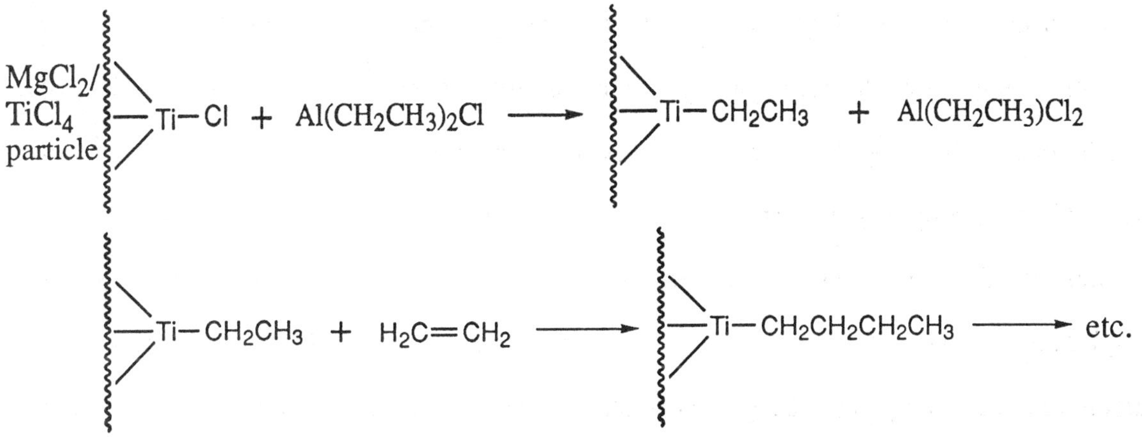

Since the initial discovery of this reaction, considerable progress has been made in understanding the mechanism involved, resulting in the development of some highly active organometallic catalysts for this transformation.

Over 60 billion pounds of polyethylene are produced world-wide every year using optimized Ziegler-Natta catalysts. Large scale reactors can yield up to 60,000 pounds of polyethylene per hour. Production of polymer at this scale is partly due to the mild conditions required for a Ziegler-Natta polymerization and is a welcome improvement over the 200 atm of pressure of ethylene and 150 to 200°C temperature necessary for the analogous free-radical methods. The more important reason, however, for the success of the Ziegler-Natta process is that the polymer obtained has substantially different physical and mechanical properties from that obtained by the free-radical route. Polyethylene from Ziegler-Natta systems, termed "high-density polyethylene" (HDPE), is three to ten times stronger than the "low-density polyethylene" (LDPE) produced in free-radical polymerizations and is opaque rather than transparent. This added strength and opacity is due to a much higher degree of crystallinity in the product as a consequence of the polymer chains being nearly linear rather than highly branched, thus allowing them to pack together far more efficiently. Because the extensive branching in LDPE prevents the polyethylene chains from packing efficiently, there is little long-range structure within these materials. LDPE is largely amorphous, with only a small amount of crystallites of a size too small to scatter light. As would be expected based on the different degrees of order in the two materials, the melting point (T_m) increases from approximately 108°C for LDPE to 133°C for HDPE. Even greater improvements in properties can be realized through special processing techniques. In the melt state, both forms of polyethylene adopt random coiled conformations (i.e., similar to that of cooked spaghetti). Engineers have developed special extrusion techniques that force the individual polymer chains of HDPE to uncoil and adopt an extended zig-zag conformation. These extended chains then align with one another to form highly crystalline materials. This elongation and alignment of chains imparts a high degree of stiffness and strength to the polymer. Simple HDPE processed in this fashion can be stronger and stiffer than steel! The *modulus* (stiffness) of processed polyethylene can be as high as 260 GPa, whereas the modulus of steel is only 210 GPa. (1 Pascal = 1 N/m^2, and 1 GPa = 145,000 psi). Similarly, these highly ordered polyethylene samples can be approximately four times stronger than steel (e.g., tensile strengths of 4.3 GPa versus 20 GPa). Because the density of polyethylene ($\approx$ 1.0 g/mL) is considerably less than that of steel (8.0 g/mL) or tungsten (19.3 g/mL), these comparisons of strength and stiffness are even more favorable if they are made on a per weight basis. Hence, even the simplest of polymers can display properties that are far superior to those of traditional engineering materials.

It is important to note that HDPE samples produced from Ziegler-Natta catalysts are not entirely free from branching. A significant chain-transfer reaction in these polymerizations is termed β-elimination that yields a metal hydride and a polymeric chain with an alkene endgroup (Scheme IV).

Scheme IV

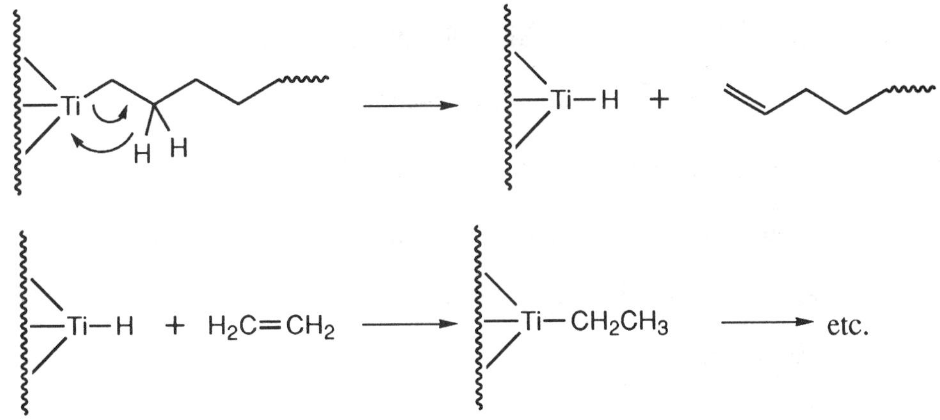

These unsaturated polymer chains then can be further polymerized to yield branched polymers (in a fashion analogous to the situation present in free-radical chain-transfer to monomers). The underlying reason why branching is rarely encountered in polymers obtained from Ziegler-Natta systems, but is common in free-radical systems, has much to do with the relative reactivities of the substituted alkenes. In free-radical reactions, substituted alkenes are more reactive than unsubstituted alkenes (ethylene); thus, once formed they immediately react and are incorporated into the polymer chains as branch points. The enhanced reactivity of substituted alkenes can be understood by comparing the relative stability of the secondary radical formed by addition to the substituted alkene with the primary radical formed from additions to ethylene. (Recall that secondary radicals are approximately 2 to 3 kcal/mol more stable than primary radicals.) In contrast, heterogeneous Ziegler-Natta catalysts exhibit a marked selectivity for reactions with ethylene rather than substituted alkenes, so even though substituted alkenes are formed during Ziegler-Natta polymerizations they are only infrequently incorporated into the polymer chains.

In recent years there here been several important advances made to the catalysts used in Ziegler-Natta type polymerizations. One of the most important has been the discovery of soluble complexes that catalyze the polymerization of ethylene or propylene at extraordinary rates.

Because these new homogeneous catalysts are substantially different in structure than the early Ziegler-Natta systems, these polymerizations are best referred to as **coordination polymerizations.** Catalysts for coordination polymerizations are frequently formed by allowing bis(cyclopentadienyl)dimethylzirconium (abbreviated $Cp_2Zr(CH_3)_2$, where Cp stands for the cyclopentadienyl anion, $C_5H_5^-$) to react with methaluminoxane (MAO). (MAO is a complex mixture of methylaluminum oxide oligomers, $[-(CH_3)AlO-)_n]$, formed by allowing trimethylaluminum to react with small amounts of water.) It is thought that MAO activates the zirconium by abstracting a methyl anion forming a zirconium cation that is the active polymerization catalyst Scheme V.

Scheme V

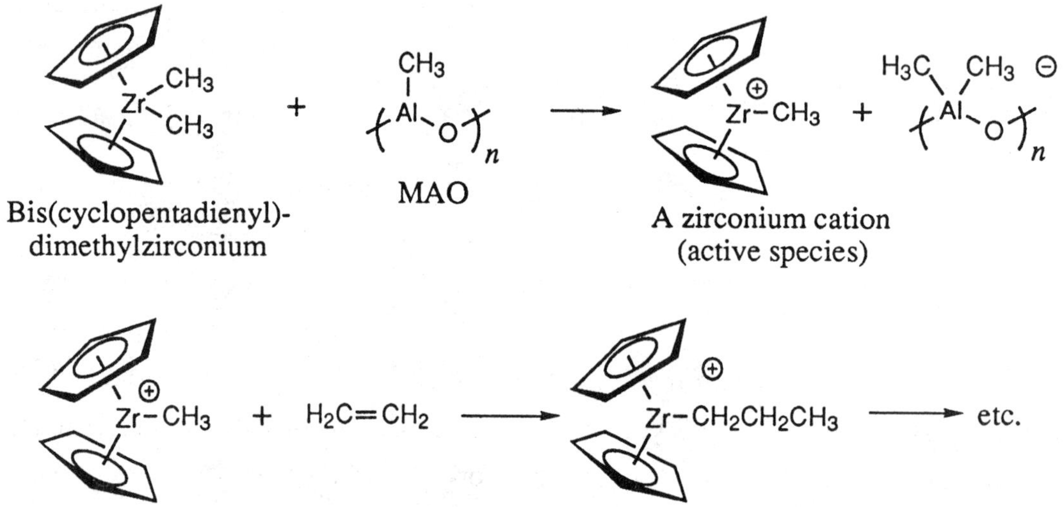

Bis(cyclopentadienyl)-
dimethylzirconium

MAO

A zirconium cation
(active species)

Some of these coordination-polymerization catalysts are very active and will polymerize up to 20,000 ethylene monomers per second, a kinetic rate typically only reached by enzyme-catalyzed biological reactions. Another important characteristic of these catalysts is that they show higher reactivity toward 1-alkenes, allowing for the formation of copolymers of ethylene and 1-hexene or 1-octene.

Copolymers of this type with these moderate length branches (C_4, C_6, and so on) are called linear-low-density-polyethylene, or LLDPE. These are useful materials because they have many of the properties of LDPE made from free-radical reactions but are formed at the substantially milder conditions associated with Ziegler-Natta polymerizations.

Chemistry in Action

The Development of Ziegler-Natta Catalysts

The discovery of Ziegler-Natta polymerization catalysts in the 1950s ranks as one of the most important scientific advances of this century. Not only are these catalysts directly responsible for a number of synthetic materials in use today, but the research efforts involved in their development elucidated many of the most basic principles of polymer chemistry and transformed this still-new discipline into an important modern science. Our current understanding of tacticity and metal-mediated polymerizations, in particular, has been made possible through the work of Karl Ziegler and Giulio Natta.

By the 1950's, polymer chemists understood that regular microstructures in polymers were necessary for strength and crystallinity but had no means to prepare macromolecules containing them. The ionic and radical initiators used at the time tended to produce atactic, amorphous polymers that had limited use as plastics or fibers. Polyethylene and polypropylene, as well, were available only as low-molecular weight, amorphous materials, and their free-radical synthesis required pressures of up to 2800 atmospheres.

In 1953 Karl Ziegler and his colleagues at the Max Planck Institute in Germany were studying the use of aluminum alkyls as possible ethylene polymerization catalysts. These compounds gave the desired product, but only with a degree of polymerization of 100 or so. At this point a competing reaction known as β-elimination began to occur. One day Ziegler performed a β-elimination polymerization experiment and obtained a simple

41

ethylene dimer instead of the expected 100-mer. After discovering that some traces of nickel in the reaction vessel left over from a previous experiment were responsible, he saw a possible solution to his problem. If nickel caused the aluminum alkyl to favor β-elimination over polymerization, would another metal have the opposite effect, leading to a higher molecular weight product?

Ziegler tested his theory by carrying out several ethylene polymerization reactions, each containing a small amount of a different transition metal salt. The polymerizations contaminated with cobalt and platinum (which are electron-rich, like nickel) also produced the dimer. However, the reaction containing Group IV transition metal zirconium gave, to his delight, "a solid cake of snow-white polyethylene," which proved to have a molecular weight higher than any previously reported. Other early metals, including titanium, exhibited similar behavior.

Ziegler disclosed his findings to the Italian chemical company Montecatini. One of their consultants, Giulio Natta, was intrigued by the new catalysts and began to study their behavior toward substituted olefins. In 1954 he was able to produce stereoregular, high-molecular weight polypropylene for the first time. After using x-ray diffraction to determine that the pendant methyl groups along the backbone were all oriented in the same direction, Natta coined the word *isotactic* to describe his new polymer (we also owe the terms *atactic* and *syndiotactic* to him). Within a short time Natta had extended the use of "Ziegler catalysts" to the polymerization of many other monomers, including α-olefins (styrene, 1-butene) and dienes (1,3-butadiene). He also elucidated many previously unknown details concerning polymer stereoregularity, and found that by choosing vanadium as the catalyst transition metal, he could produce syndiotactic polypropylene.

Uses for the new polymers made with Ziegler-Natta catalysts developed immediately. Montecatini began producing isotactic polypropylene commercially in 1957. Its lightweight properties, combined with the low cost of the monomer, have made it one of the most popular fibers in use today. Scientists at the Gulf-Goodrich company, which had also received a disclosure from Ziegler in 1953, found that the microstructures of the natural isoprene rubbers gutta-percha (*trans*-1,4) and Hevea (*cis*-1,4) could be duplicated with Ziegler-Natta polymerization catalysts. These "natural" synthetic rubbers, along with the 1,4-polybutadiene first made by Natta, are undoubtedly of great importance to our society. Polyacetylene, which is under investigation as a potential electrical conductor, is just one of the many other useful polymers made by this method. For their important contributions to science, Ziegler and Natta were awarded a joint Nobel Prize in chemistry in 1963.

5.D. Ionic Chain-Growth Polymerizations.

Chain-growth polymers can also be synthesized using reactions that rely on either anionic or cationic species in the propagation steps. The choice of ionic procedure depends greatly on the electronic nature of the monomers to be polymerized. Vinyl monomers with electron-withdrawing groups, which stabilize carbanions, are used in anionic polymerizations, whereas vinyl monomers with electron-donating groups are used for cationic polymerizations.

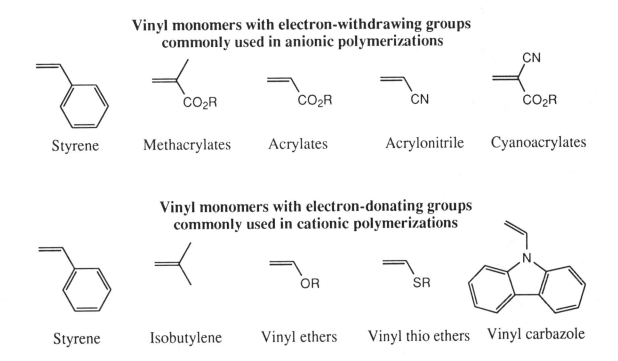

Vinyl monomers with electron-withdrawing groups commonly used in anionic polymerizations

Styrene   Methacrylates   Acrylates   Acrylonitrile   Cyanoacrylates

Vinyl monomers with electron-donating groups commonly used in cationic polymerizations

Styrene   Isobutylene   Vinyl ethers   Vinyl thio ethers   Vinyl carbazole

Styrene is conspicuous among the above monomers in that it can be polymerized using either anionic or cationic techniques. This ambidextrous characteristic is due to the phenyl ring, which can act to stabilize either a cation or an anion through resonance. Note that the propagating endgroup is benzylic in an anion or a cation in a styrene polymerization.

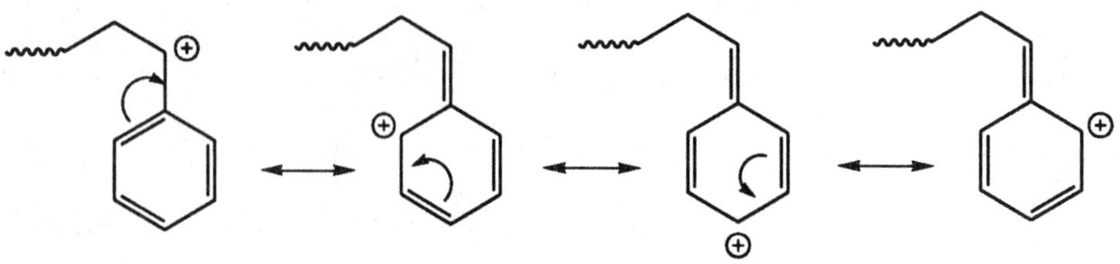

Stabilization of a benzylic cation endgroup
through resonance

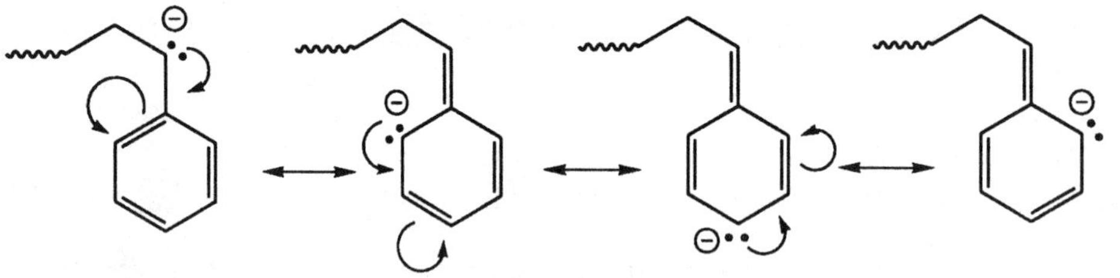

Stabilization of a benzylic anion endgroup
through resonance

Scrupulous purity of reagents and solvents is far more important in ionic polymerizations than in the free-radical polymerizations. Carbanion chain ends are typically conjugate bases of very weak acids and, they are strong bases that are easily protonated by water.

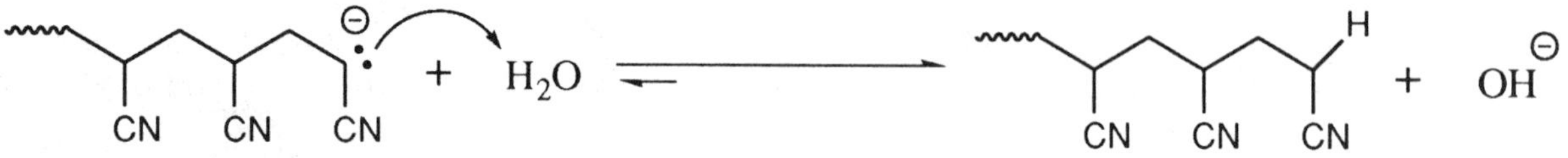

| Stronger base | Stronger acid | | Weaker acid | Weaker base |

Similarly, carbocations can act as very strong acids and deactivate by losing a proton, or alternatively they act as Lewis acids and deactivate by reaction with a Lewis base.

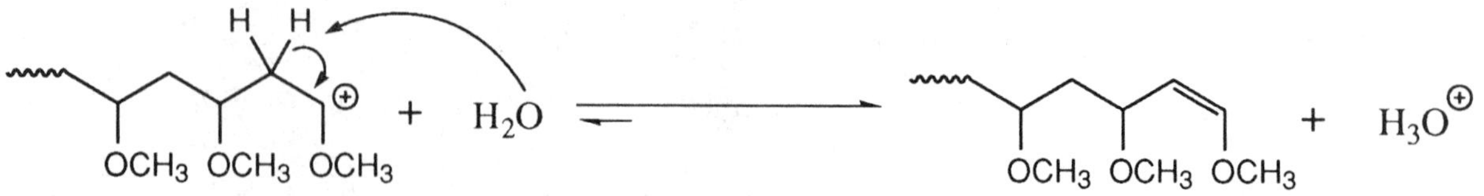

| Stronger acid | Stronger base | | Weaker base | Weaker acid |

44

Carbanions and carbocations are also deactivated by reactions oxygen that involve redox processes. As a result, water and oxygen are ubiquitous impurities that must be meticulously eliminated from these reactions. The reaction temperature range for ionic polymerizations is considerably different from that used in other polymerizations. Step-growth condensations and free-radical reactions are typically run well above room temperature. In contrast, both anionic and cationic polymerizations are run at very low temperatures ($\approx -78°C$), in hopes of averting unwanted termination and transfer reactions.

Anionic Polymerizations. Anionic polymerizations can be initiated either by addition of a nucleophile to an activated alkene or by means of an electron-transfer process. The former approach typically involves the use of metal alkyls such as methyl or *sec*-butyllithium as the nucleophilic initiating agent (equation 24).

$$\text{etc.} \quad (24)$$

The newly formed carbanion then acts as a nucleophile and adds to another equivalent of monomer, and the propagation continues.

Under anionic conditions, short chains of polyethylene can be formed but true polymerization is difficult because the chains precipitate before substantial growth can occur. Propylene and other monomers with allylic protons undergo rapid deprotonation with organolithium reagents to form stable allylic anions that are not reactive enough to add to another equivalent of monomer and thus propagate chain growth (equation 25).

$$CH_3Li \quad + \quad \longrightarrow \quad CH_4 \quad + \quad \longrightarrow \quad (25)$$

Viable substrates for anionic polymerizations, therefore, do not have α-protons.

Another method in widespread use for the initiation of anionic polymerizations involves one-electron reduction of the monomer to form a radical anion. The radical anion thus formed can then either be further reduced to form a dianion (equation 26) or can dimerize, again yielding a dianion (equation 27).

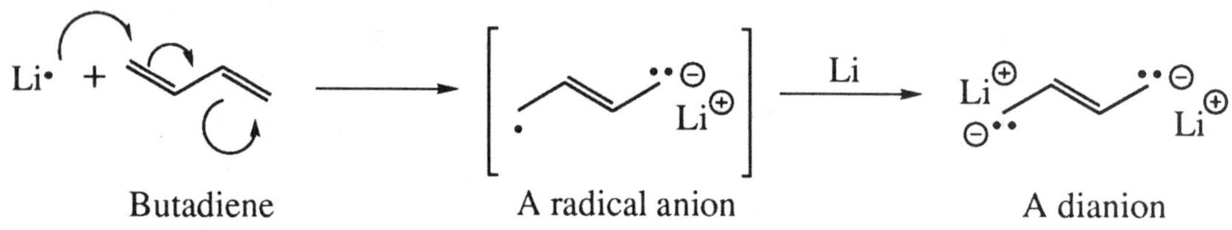

Butadiene   A radical anion   A dianion   (26)

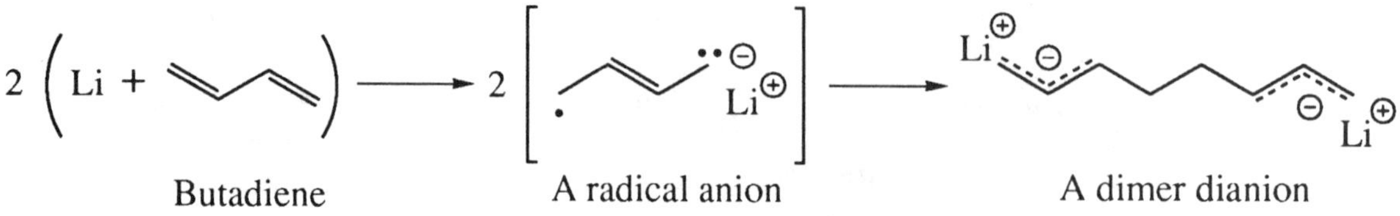

Butadiene   A radical anion   A dimer dianion   (27)

In either case, a single initiator can now propagate chains from both ends, by virtue of its two active endgroup carbanions. Group IA metals such as Li or Na can be used as reducing agents. These reactions are heterogeneous and involve transfer of the electron at the surface of the metal. In order to improve the efficiency of this process, soluble reducing agents such as sodium naphthalide can be used. Sodium undergoes electron-transfer reactions with extended aromatic π-systems such as naphthalene to form soluble radical anions (equation 28).

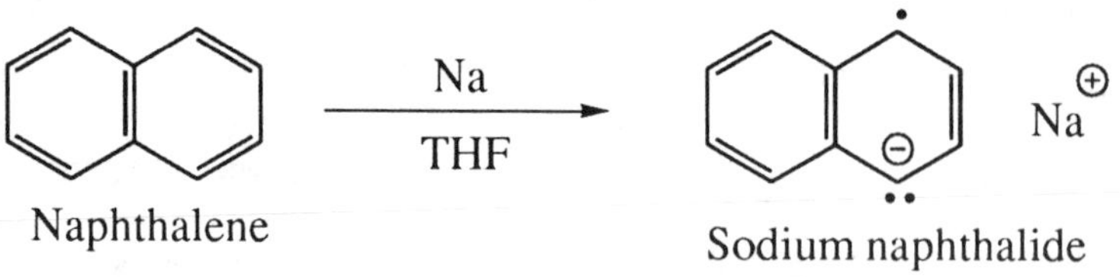

Naphthalene   Sodium naphthalide   (28)

The naphthalide radical anion is a powerful reducing agent. For example, styrene undergoes a one-electron reduction to form the styryl radical anion, which then couples to form the dianion as described previously. The latter then propagates polymerization at both ends, growing chains in two directions simultaneously (Scheme VI).

46

Scheme VI

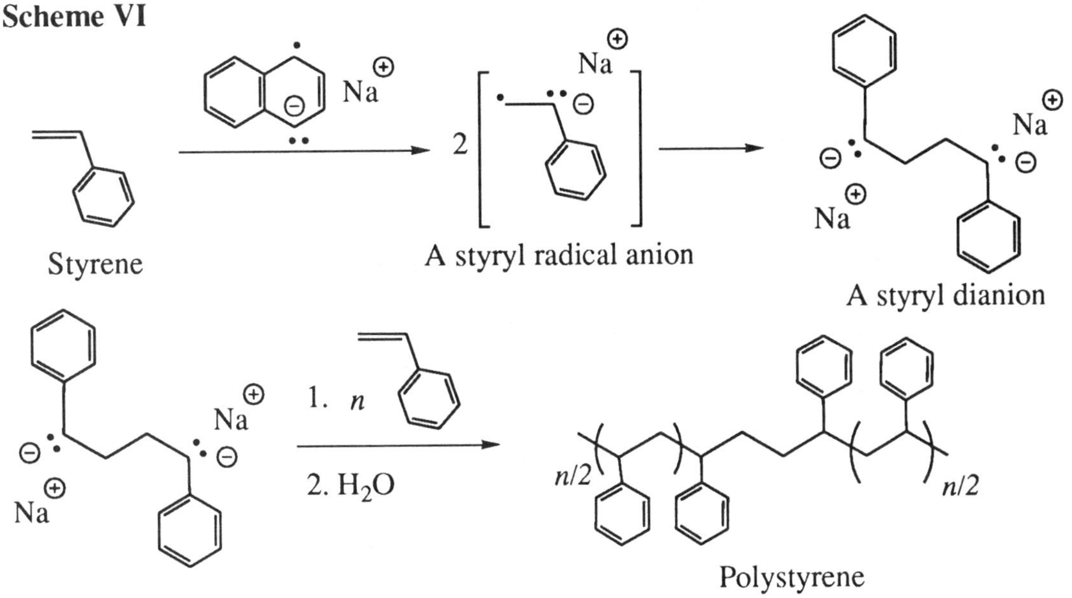

Styrene → A styryl radical anion → A styryl dianion

1. n [styrene]
2. H_2O

→ Polystyrene

The propagation characteristics of anionic polymerizations are similar to those of free-radical polymerizations, but with the important distinction that many of the chain-transfer and termination reactions that plague radical processes are absent. The use of solvents and monomers that do not have acidic protons can eliminate many of the chain-transfer reactions. Furthermore, because the propagating chain ends carry the same charge, bimolecular coupling and disproportionation reactions are also averted. An interesting set of circumstances arises when chain-transfer and chain-termination steps are no longer significant. Under these conditions, polymer chains are initiated and continue to grow without abatement until either all of the monomer is consumed or some external agent is added to terminate the chains. Polymerizations of this type are termed **living polymerizations** and figure prominently in the preparation of specialty polymers.

The absence of chain-transfer and chain-termination steps in living polymerizations has far-reaching consequences. One of the most visible of these consequences is in the area of molecular weight control. The molecular weight of a polymer originating from living polymerizations is determined directly by the monomer-to-initiator ratio. It is, therefore, relatively easy to obtain polymers of a defined size simply by controlling the stoichiometry of the reagents. In contrast, the average sizes of polymer chains formed from nonliving, chain growth

47

processes (free-radical, Ziegler-Natta, and so on) vary from system to system and are determined by the ratio of the rate of propagation to the rate of termination. In most cases, precise control over the molecular weight of the product obtained in nonliving systems is not possible, because it is very difficult to change one of the rates involved without affecting the other.

Living polymerizations also provide a second important refinement over nonliving polymerizations: the ability to prepare polymers with narrow molecular-weight distributions. Under ideal conditions, living systems yield polymers with polydispersity indices of 1.1 or less, whereas nonliving chain-growth processes typically have much broader molecular-weight distributions, with polydispersities of 2.0 or greater.

The well-controlled character of living polymerizations can be used to help probe mechanistic details of a reaction. For example, the reactivity of carbanions is highly dependent upon a variety of factors including solvent, temperature, and the nature of the counterions, because of the complex equilibria of species that are present in solution of ionic complexes. The four species that can be involved in these systems are shown in equation 29.

$$R-M \rightleftharpoons R^{\ominus} M^{\oplus} \rightleftharpoons R^{\ominus} \| M^{\oplus} \rightleftharpoons R^{\ominus} + M^{\oplus} \quad (29)$$

$$\begin{array}{cccc} \text{Covalent} & \text{Contact} & \text{Solvent separated} & \text{Free ions} \\ & \text{ion pair} & \text{ion pair} & \end{array}$$

M = Metal ion

The position of this equilibrium will shift depending upon the specifics of the system, but in general, at least two of these species will be present. The activity of these individual species increases in going from left to right as written, with the solvent-separated ion pair and the free ions being the most important. As chain ends, the free ion adds monomer to the growing chain faster than the solvent-separated chain ends (i.e., $k_{free} > k_{ss}$ in Scheme VII). At the same time, these ionic endgroups interconvert with rate constants k_1 and k_{-1}.

Scheme VII

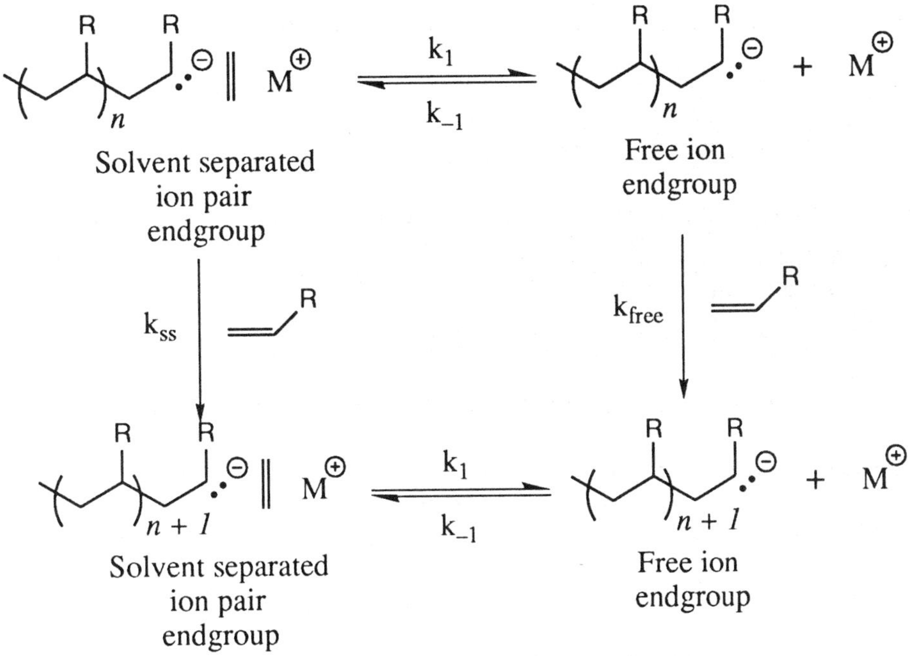

Because living polymerizations are expected to yield polymers with narrow molecular-weight distributions, it is possible to use the breath and shape of the distribution curve to yield information about the relative rates of these different processes. If exchange between ionic endgroups is very slow relative to the rate of propagation, then the two different types of chains (free ion endgroups and solvent separated ion endgroups) propagate independently of one another and a bimodal distribution of molecular weights results (i.e., a molecular-weight distribution that shows two distinct maxima). The higher molecular-weight peak corresponds to the free ion chains that propagate faster, and add more monomer in comparison to the solvent separated ion chains (Figure 4). In contrast, if the rate of exchange is very fast relative to the propagation rate, then the polymerization propagates as if their were only one type of chain end present and a narrow, monomodal distribution results. Polymers obtained from reactions in which the rates of exchange are comparable to the rates of propagation have a broadened distribution (Figure 4).

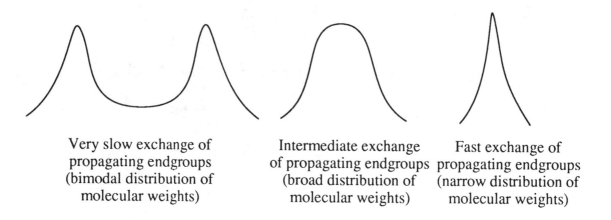

Very slow exchange of propagating endgroups (bimodal distribution of molecular weights)

Intermediate exchange of propagating endgroups (broad distribution of molecular weights)

Fast exchange of propagating endgroups (narrow distribution of molecular weights)

Figure 4. Molecular-weight distributions from living polymerizations having different relative rates of exchange and propagation.

After consumption of the monomer under living, anionic conditions, electrophilic terminating agents can be added to functionalize the chain ends. Examples of terminating reagents include CO_2 and ethylene oxide, which, after protonation, form carboxylic acids and alcohol-terminated chains, respectively (equation 30).

(30)

In a similar fashion, polymer chains with functional groups at both ends, called **telechelic polymers,** can be prepared by addition of these same reagents (CO_2, ethylene oxide, and so on) to solutions of chains with two active ends initiated by sodium naphthalide.

Example 7

Show how to prepare polystyrene that is terminated at both ends with carboxylate groups.

Solution

The most straightforward pathway involves the formation of growing chains with two active endgroups by the treatment of styrene by sodium naphthalide.

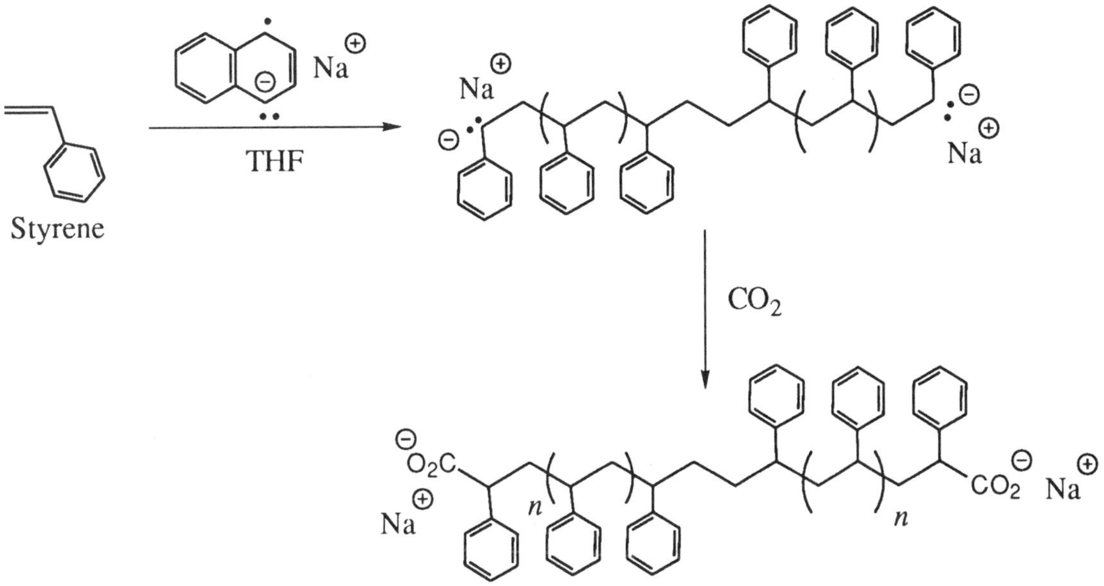

A telechelic polystyrene dicarboxylate

Problem 6

Show how to prepare polybutadiene that is terminated at both ends with hydroxyl groups.

Specific chain-end functionalization is not possible in nonliving polymerizations such as the free-radical systems discussed previously because chains are initiated by the free-radical initiators as well as through chain-transfer reactions. In addition, chain-termination occurs through the bimolecular processes of coupling and disproportionation. The result is a disparate collection of polymer chains with various combinations of endgroups, representing all of these possibilities. Living polymerizations are initiated in one way exclusively and are terminated in one way exclusively.

Because chain ends are still active after consumption of the monomer in living polymerizations, a second portion of monomer can be added to the reaction vessel and polymerization will resume (equation 31).

$$R-Li \xrightarrow{\quad n \quad} \quad \xrightarrow{\quad m \quad} \quad \tag{31}$$

Polymers formed by the sequential addition of different monomers are called **diblock copolymers**, and similarly, addition of a third monomer leads to the formation of **triblock copolymers**. If the two blocks are of sufficient molecular weight, the diblock copolymer will retain the properties of each of the two homopolymers. More important, diblock and triblock copolymers exhibit unique properties that are not found in either the homopolymers or in random copolymers. This useful phenomenon arises from the composite-like morphology adopted by these block copolymers. Individual blocks within the polymer chains tend to phase separate into their own discrete domains. Examples are diblocks and triblocks possessing polybutadiene and polystyrene segments.

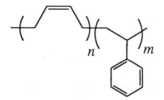

A 1,4-*cis*-butadiene-styrene
diblock copolymer

A styrene-1,4-*cis*-butadiene-styrene
triblock copolymer

Pure poly(1,4-*cis*-butadiene), $T_g = -102°C$, is an amorphous rubber polymer at room temperature, whereas pure polystyrene is a stiff, brittle, glassy polymer with a $T_g = 100°C$. In contrast, diblock copolymers having both polybutadiene and polystyrene segments are strong, somewhat flexible, and very tough. The styrene-butadiene-styrene triblock copolymer constructed with the amorphous polybutadiene segment sandwiched between two glassy polystyrene segments acts as an elastomer. Recall that an elastomer is a low T_g polymer crosslinked so that the chains do not irreversibly slip past one another during deformation. In the styrene-butadiene-styrene triblock, the glassy polystyrene segments act as physical crosslinks connected by the amorphous polybutadiene segments.

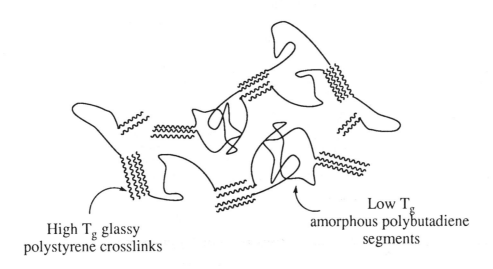

High T_g glassy
polystyrene crosslinks

Low T_g
amorphous polybutadiene
segments

Phase separated morphology of styrene-butadiene-styrene
triblock copolymers

A crosslinked network is formed when the two polystyrene segments on a single chain solidify into different glassy domains. Although the simple diblock copolymer also phase separates into discrete rubbery and glassy domains, these domains are not tied together because the polymer chains lack the essential second polystyrene block. Diblocks are therefore incapable of forming a similar network. Because the triblock's crosslinks are physical and not chemical, they break apart when the material is heated above the polystyrene's T_g. Elastomers of this type are called *thermoplastic elastomers* and are useful because they can be recycled.

Cationic Polymerizations. Cations also can be used as the propagating species in chain-growth polymerizations. Like carbanions, carbocations are highly reactive species subject to a variety of unwanted side reactions; therefore, low temperatures and highly purified conditions must be maintained in running these reactions. Only alkenes with electron-donating substituents are active toward cationic polymerizations. Examples of effective electron donating groups include alkyl, aryl, ether, thioether, and amino groups.

Although the initiating carbocations can be generated in several ways, two approaches are most common. The first approach involves the addition of a strong protic acid to an organic monomer or, alternatively, the abstraction of a halide from the organic initiator by a Lewis acid.

Initiation by protonation of an alkene requires the use of a strong acid with a non-nucleophilic counterion in order to avoid 1,2-addition across the alkene double bond (equation 32).

(32)

Suitable non-nucleophilic counter ions include SO_4^{2-}, AsF_6^-, BF_4^-, and $B(C_6H_5)_4^-$.

The second common method for generating carbocations involves the reaction between alkyl halide co-initiators and strong Lewis acids such as BF_3, $SnCl_4$, $AlCl_3$, $Al(CH_3)_2Cl$, and $ZnCl_2$. With trace water present, the mechanism of initiation using some Lewis acids is thought to involve protonation of the alkene (equation 33).

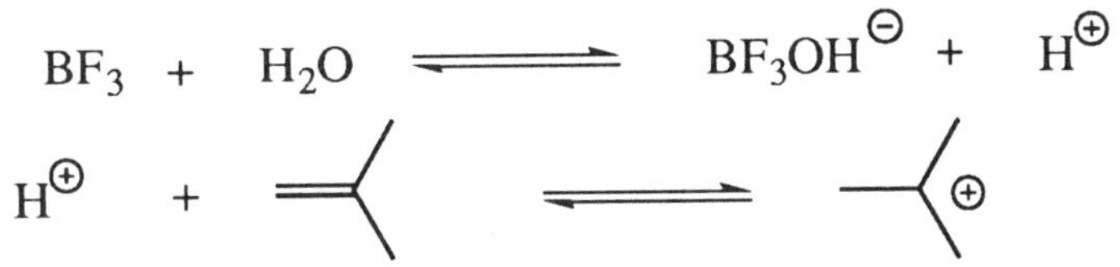

$$BF_3 + H_2O \rightleftharpoons BF_3OH^{\ominus} + H^{\oplus} \qquad (33)$$

$$H^{\oplus} + \text{2-Methylpropylene} \rightleftharpoons$$

2-Methylpropylene

In the absence of water, the Lewis acid coordinates to and removes the halide from the organic co-initiator to form the initiating carbocation (equation 34).

$$\text{2-Chloro-2-phenylpropane} + SnCl_4 \rightleftharpoons + SnCl_5^{\ominus} \qquad (34)$$

2-Chloro-2-phenylpropane

The polymerization of alkenes then propagates by the electrophilic attack of the carbocation on the double bond. The regiochemistry of the addition is directed by the formation of the more stable (i.e., the more highly substituted) carbocation. The relative stabilities of carbocations ($3° > 2° > 1°$) is important in determining the course of these reactions, because these reactive species will either react or rearrange to yield a more stable carbocation if possible. Terminal alkenes are relatively electron-rich alkenes but are surprisingly poor substrates for cationic reactions. The reason for their low reactivity is thought to be related to a hydride transfer reaction between the 2° carbocationic chain end and a hydrogen along the polymer backbone, to yield a new 3° carbocation (equation 35).

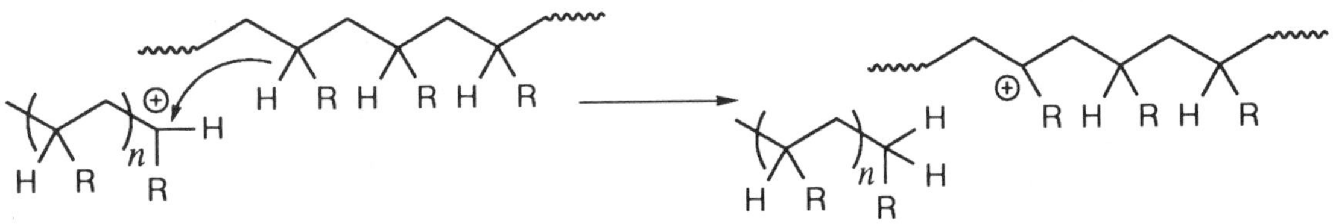

$$\qquad (35)$$

The newly formed 3° carbocation is approximately 12 to 15 kcal/mol more stable than the 2° carbocation that would be formed by addition of another equivalent of monomer. Hence, this hydride transfer reaction, like any other reaction that would generate a more stable carbocation, effectively terminates the polymerization.

Example 8

Write a mechanism for the polymerization of isobutylene (2-methylpropene) initiated by 2-chloro-2-phenylpropane and $SnCl_4$. Label the initiation, propagation, and termination steps.

Solution

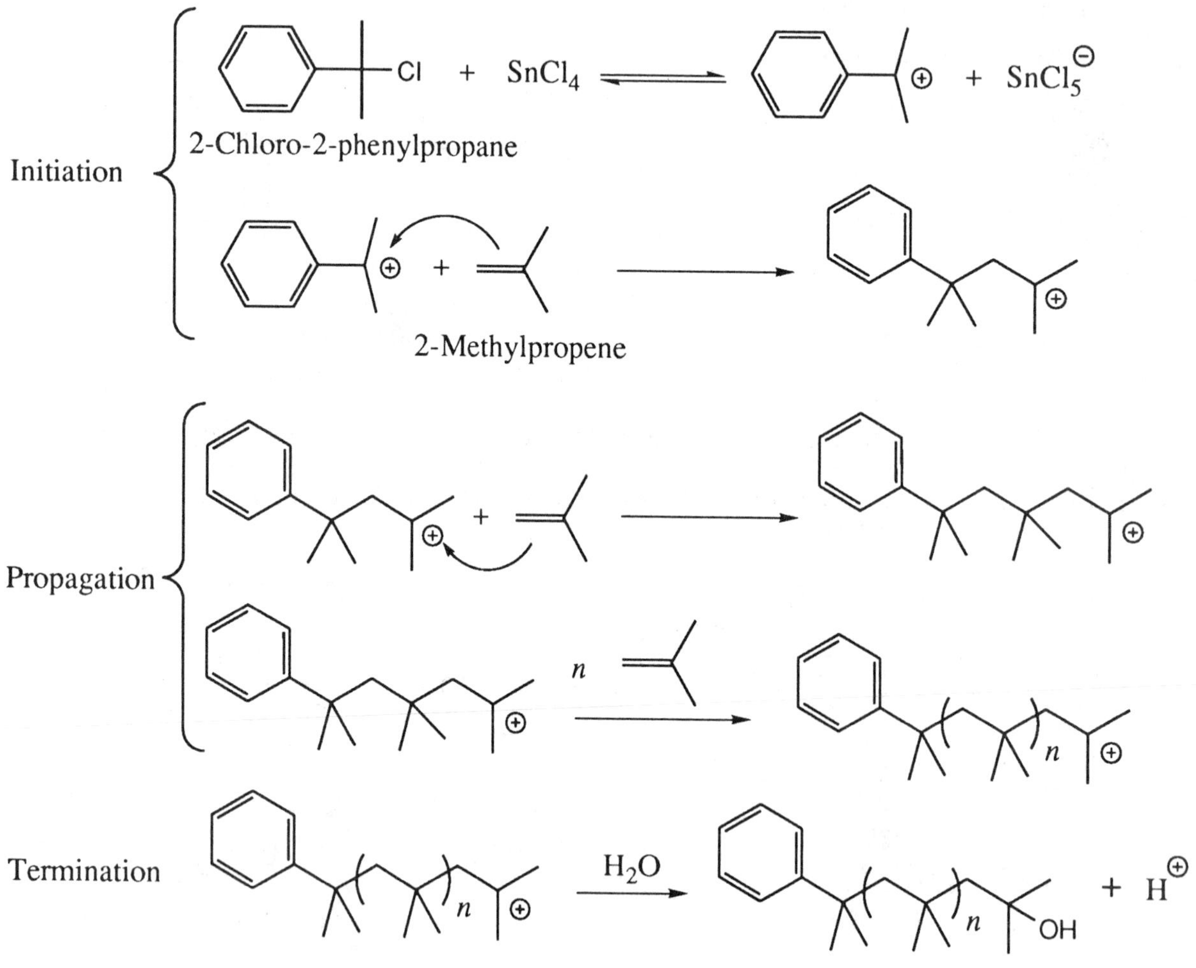

Problem 7

Write a reasonable mechanism for the polymerization of methyl vinyl ether initiated by 2-chloro-2-phenylpropane and SnCl4. Label the initiation, propagation, and termination steps.

In cationic polymerizations, chain-transfer reactions can occur by loss of a proton from the propagating chain end. If the base that accepts the proton is the monomer, then a new carbocation is formed, initiating the growth of another chain (equation 36).

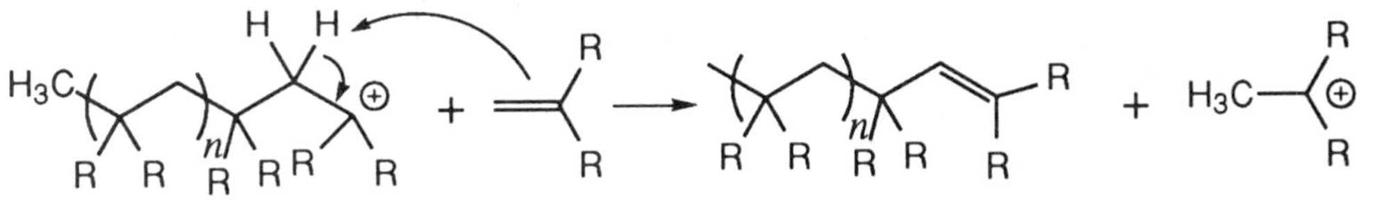

(36)

Strained cyclic molecules such as ethers and epoxides can also be polymerized under acidic conditions. For example, THF undergoes ring-opening polymerization by treatment with strong acids having non-nucleophilic counterions, such as perchloric acid. The polymerization propagates through an S_N2 attack on the α-carbon of an oxonium ion (equation 37).

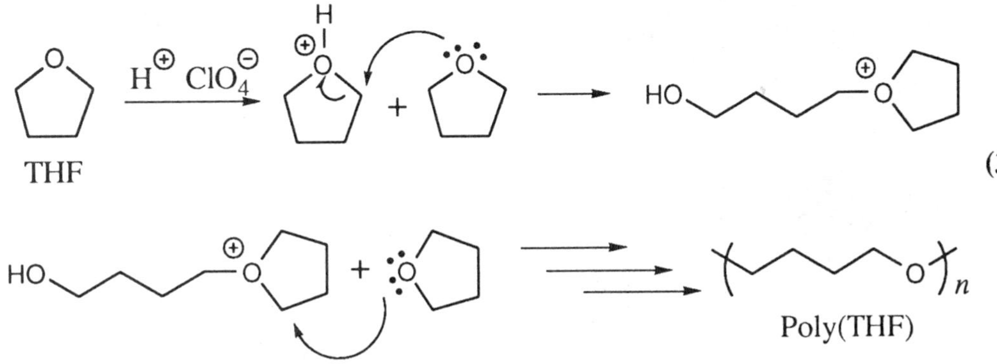

THF

(37)

Poly(THF)

5.E. Ring-Opening Metathesis Polymerizations

During early investigations into the polymerizations of cycloalkenes by transition metal catalysts (i.e., Ziegler-Natta type systems), polymers with unexpected structures were formed that retained the same number of double bonds as originally present in the monomer. This process is illustrated by the polymerization of bicyclo[2.2.1]hept-2-ene (norbornene) by $TiCl_4/LiAl(n-C_7H_{15})_4$ in equation 38.

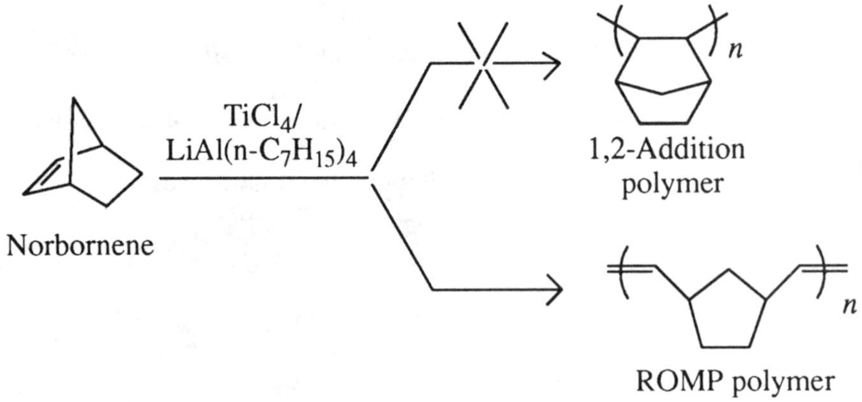

Norbornene
$TiCl_4/$
$LiAl(n-C_7H_{15})_4$
1,2-Addition polymer
ROMP polymer

(38)

The fact that the resulting polymer is unsaturated requires that this polymerization proceed by a mechanism substantially different from that involved in the polymerization of alkenes by these same catalyst mixtures. Following lengthy and detailed studies, the mechanism of this process has been revealed to proceed via a metallacyclobutane species, arising from the reaction between a metal alkene and a cycloalkene (equation 39). The intermediate metallacycle then undergoes a retro [2+2] reaction to yield a new, substituted carbene. Repetition of these steps leads to the formation of the unsaturated polymer.

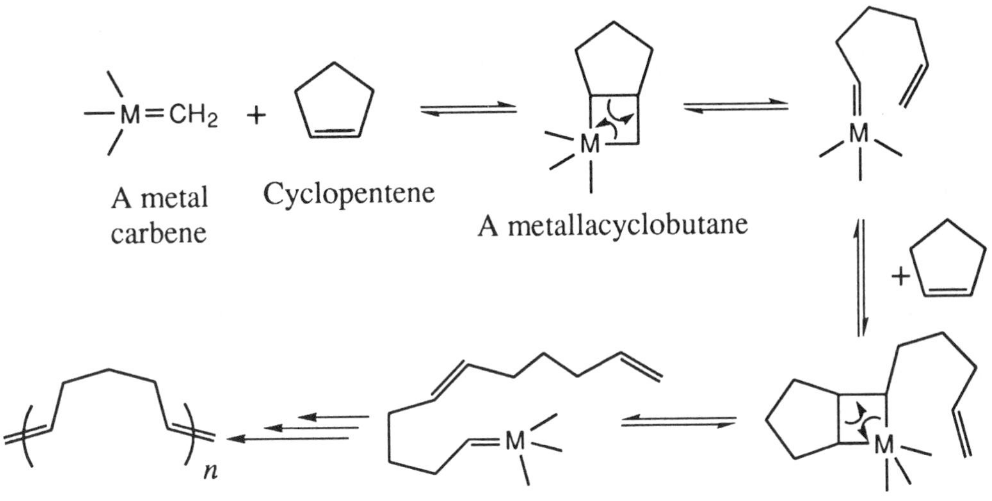

(39)

This process has been termed **ring-opening metathesis polymerization**, or **ROMP**, after a related process involving the reaction between acyclic alkenes and metal carbenes that effectively exchanges the ends of two olefins (equation 40).

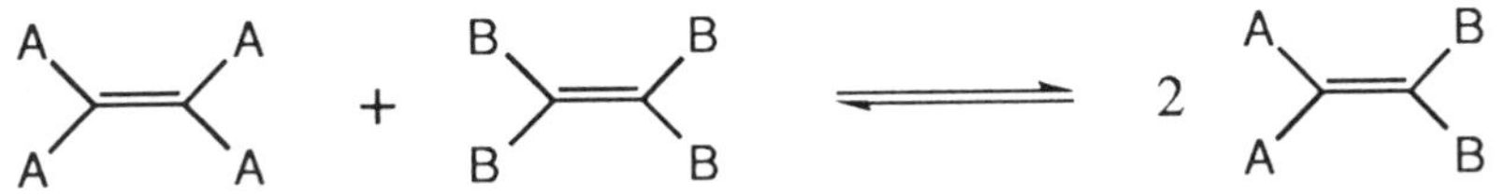

(40)

All of the steps in the ROMP process are reversible, and the reaction is driven in the forward direction by the release of angle strain that accompanies the opening of the rings. For this reason, strained cycloalkenes are often better substrates than unstrained cycloalkenes. For example, the reactivity of rings towards ROMP decreases in the order shown:

Ring Strain: 27   29.8   5.9   1.4
(kcal/mol)

Improved understanding of the mechanism involved in these transformations has led to the isolation and subsequent development of several well-defined metal carbene and metallacyclobutane complexes (shown here), all of which catalyze the ROMP reaction:

59

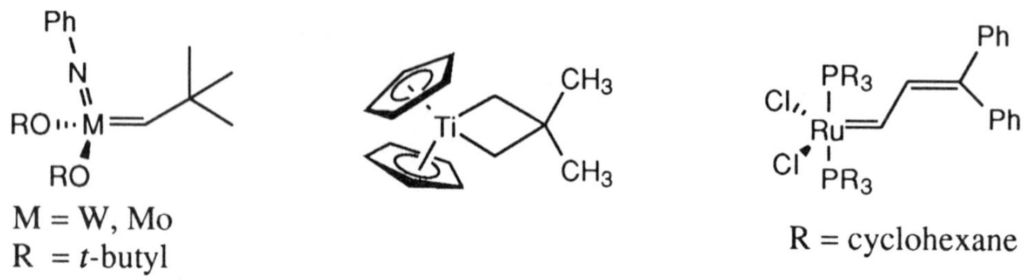

M = W, Mo
R = *t*-butyl

R = cyclohexane

Under suitable conditions these complexes may be used to initiate the living polymerization of a number of cycloalkenes. The polymerization can be terminated by taking advantage of the reactivities of the complexes with other functional groups. For example, Mo and W carbene endgroups react with aldehydes in a Wittig fashion (Section 17.8) to yield a metal oxo complex and an alkene terminated polymer. This reaction can be used as a terminating process in living polymerizations catalyzed by these complexes (equation 41).

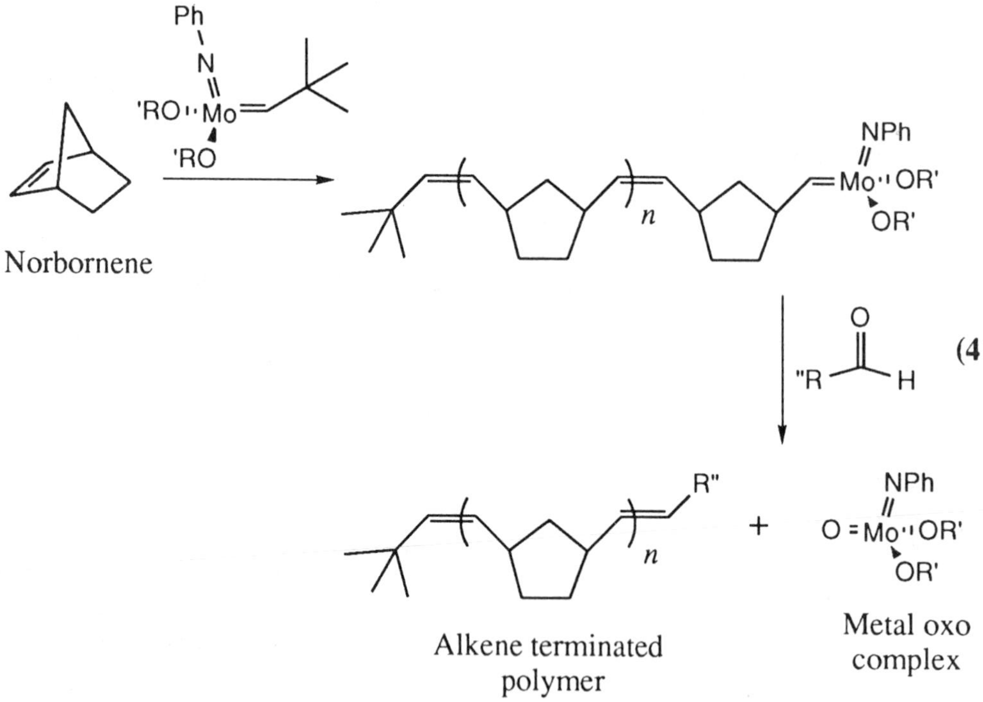

Norbornene

(41)

Alkene terminated
polymer

Metal oxo
complex

5.F. Conjugated Polymers Through ROMP Techniques.

Because of their unparalleled versatility, polymeric materials continue to maintain a high profile in new technologies. Fully conjugated polymers, typified by polyacetylene (equation 42), play an important role in emerging electronic and optical applications.

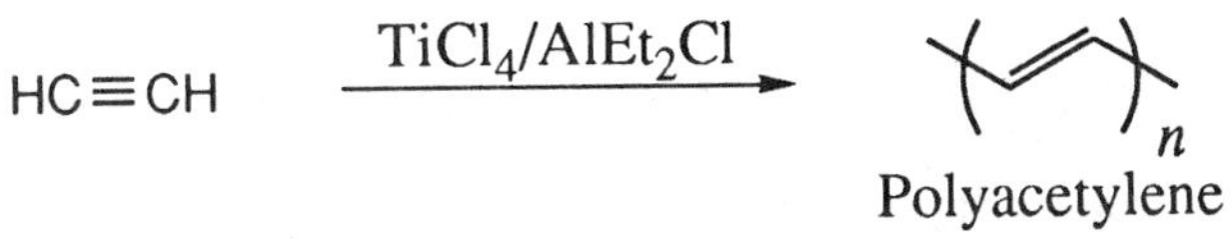

$$HC \equiv CH \xrightarrow{\text{TiCl}_4/\text{AlEt}_2\text{Cl}} \text{[Polyacetylene structure]}_n \qquad (42)$$

Polyacetylene

Polyacetylene, with its extensive conjugation, is a shiny black solid. It is insoluble in all solvents, and decomposes before melting. Its loosely held, delocalized π-electrons act much like the electrons in a metal. Not surprisingly, this material behaves as a weak electrical conductor. Upon oxidation with I_2 (Figure 5) or AsF_5 (to form radical cations), or upon reduction with sodium naphthalide anion (to form radical anions), the conductivity of polyacetylene can be increased to a value that is greater than that of mercury metal and nearly as high as that of copper metal. (Sometimes these redox treatments are referred to as "doping", which is a term that comes from the semiconductor industry.)

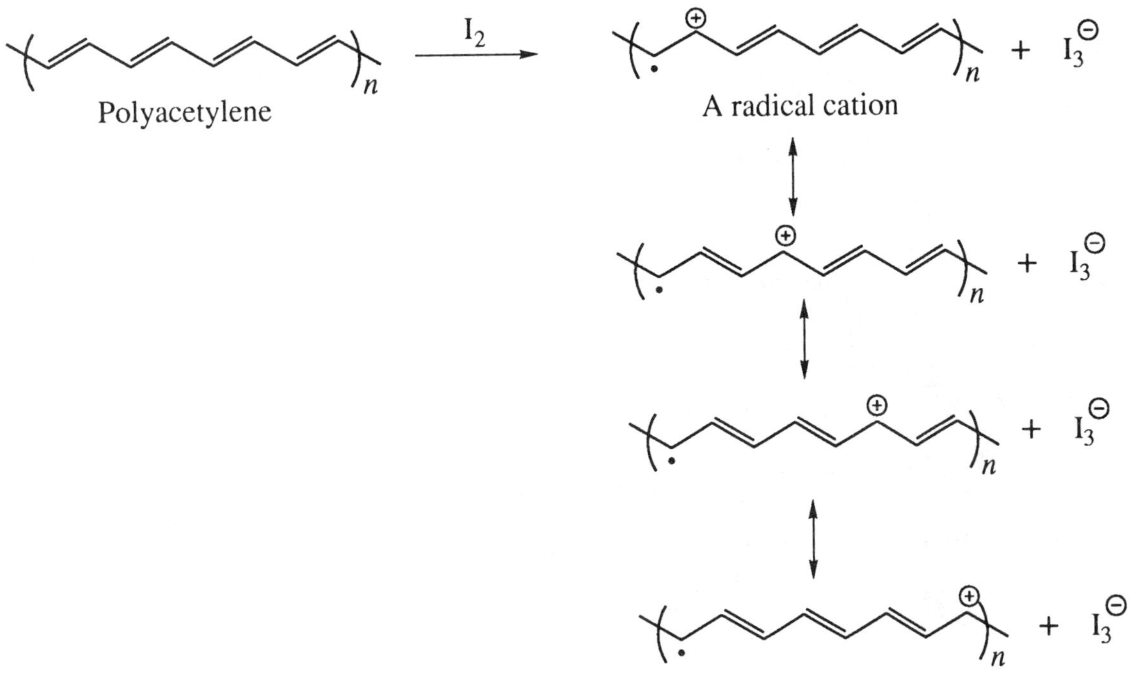

Figure 5: Doping (or oxidizing) polyacetylene using I_2 to form a radical cation that is delocalized along the polymer backbone.

61

This conductivity, which is a highly unusual property for an organic polymer, results from the extensive backbone conjugation that allows the radical cation or radical anion to delocalize along the chain (Figure 5). The movement of these charged-defect centers in a potential field constitutes a current flow. In recent years, the electronic properties of a number of fully conjugated polymers have been investigated. The structures of some of the more common conducting polymers are shown here along with their associated conductivity values in units of S/cm:

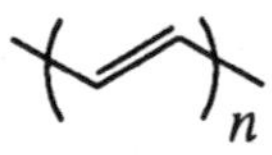

Polyacetylene
10^5 S/cm

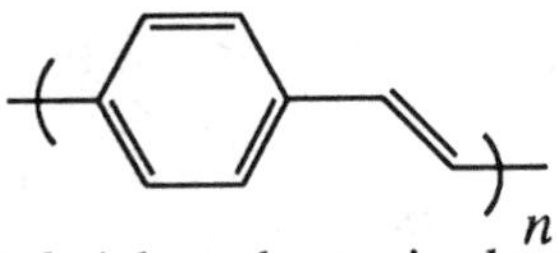

Poly(phenylene vinylene)
10^3 S/cm

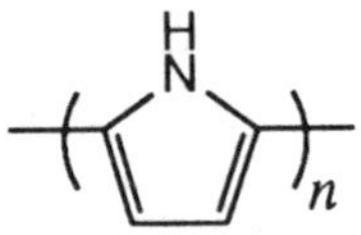

Polypyrrole
10 - 100 S/cm

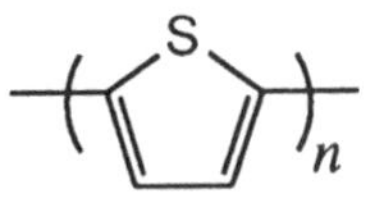

Polythiophene
10^2 S/cm

Poly(p-phenylene)
10^3 S/cm

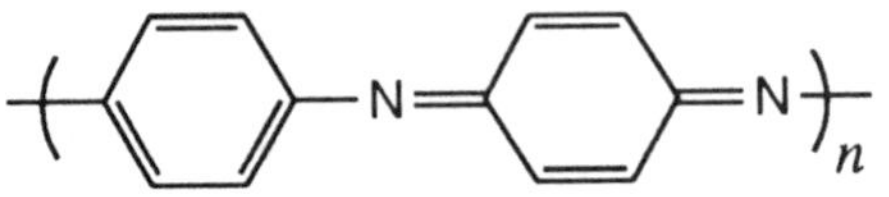

Polyaniline
10^2 S/cm

Typical metals show conductivities above 10^2 S/cm (e.g., copper $\approx 10^6$ S/cm, mercury $\approx 10^4$ S/cm); semiconductors between 10^2 and 10^{-6} S/cm (e.g., pure silicon $\approx 10^{-4}$ S/cm, germanium $\approx 10^{-1}$ S/cm), and insulators have conductivity values below 10^{-6} S/cm (e.g., polystyrene $\approx 10^{-18}$ S/cm, polyethylene $\approx 10^{-13}$ S/cm).

ROMP reactions are unique in that all of the unsaturation present in the monomers is conserved in the polymeric product. This feature makes ROMP techniques attractive for the preparation of highly unsaturated and fully conjugated materials. One example is the direct preparation of polyacetylene by the ROMP through one of the double bonds of cyclooctatetraene (equation 43).

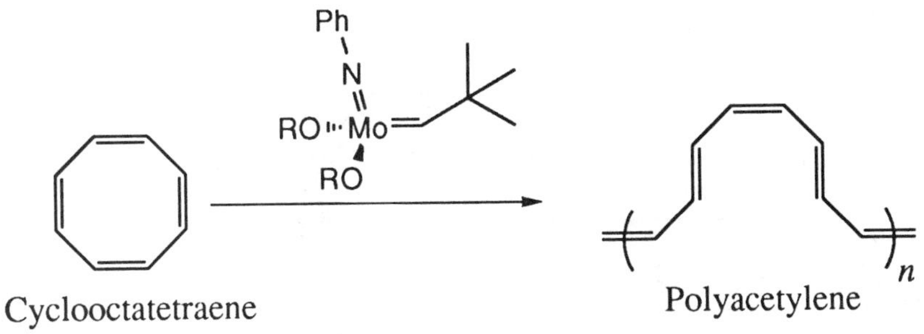

Cyclooctatetraene → Polyacetylene (43)

Alternative ROMP precursor routes into polyacetylene have also been developed. One particularly imaginative process is based on the ROMP of benzvalene (a highly strained isomer of benzene, C_6H_6) to form a polymer containing double bonds separated by bicyclo[1.1.0]butane units. Mild heating or the addition of Lewis acids results in a bicyclo[1.1.0]butane to butadiene rearrangement and complete the conjugation along the polymer backbone (equation 44).

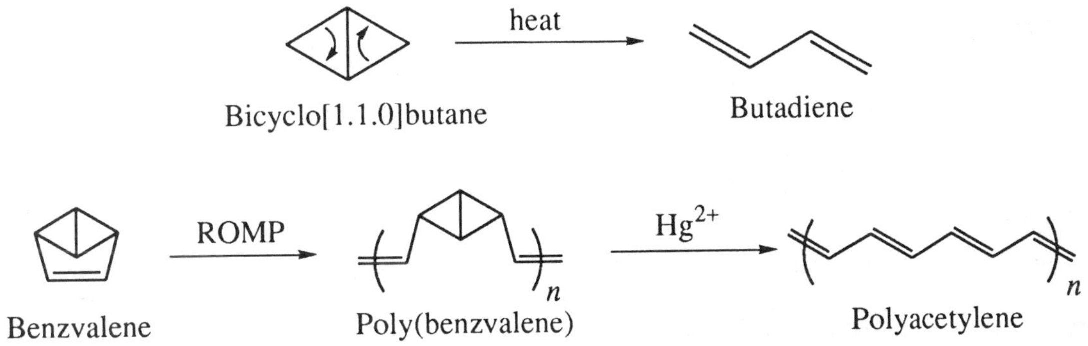

Bicyclo[1.1.0]butane → Butadiene

(44)

Benzvalene → Poly(benzvalene) → Polyacetylene

The precursor polymer, poly(benzvalene), is completely soluble in a number of organic solvents, so that it can easily be processed into thin sheets or drawn into wires before its final conversion to polyacetylene. One significant drawback to the commercialization of this last process is the fact that the both the monomer and the resulting polymer are highly explosive. To circumvent this unfortunate limitation, a non-ROMP precursor route has been developed that involves the formation of a polymer that can undergo retro-Diels-Alder reactions (retro-[4+2]) that introduce carbon-carbon double bonds in the polymer chain. Diethyl 7-oxanorbornadiene-2,3-dicarboxylate can be polymerized across the unsubstituted double bond using palladium(II) complexes. Upon heating, this soluble precursor polymer looses dimethyl furan-3,4-dicarboxylate through a retro-Diels-Alder reaction to yield polyacetylene (equation 45).

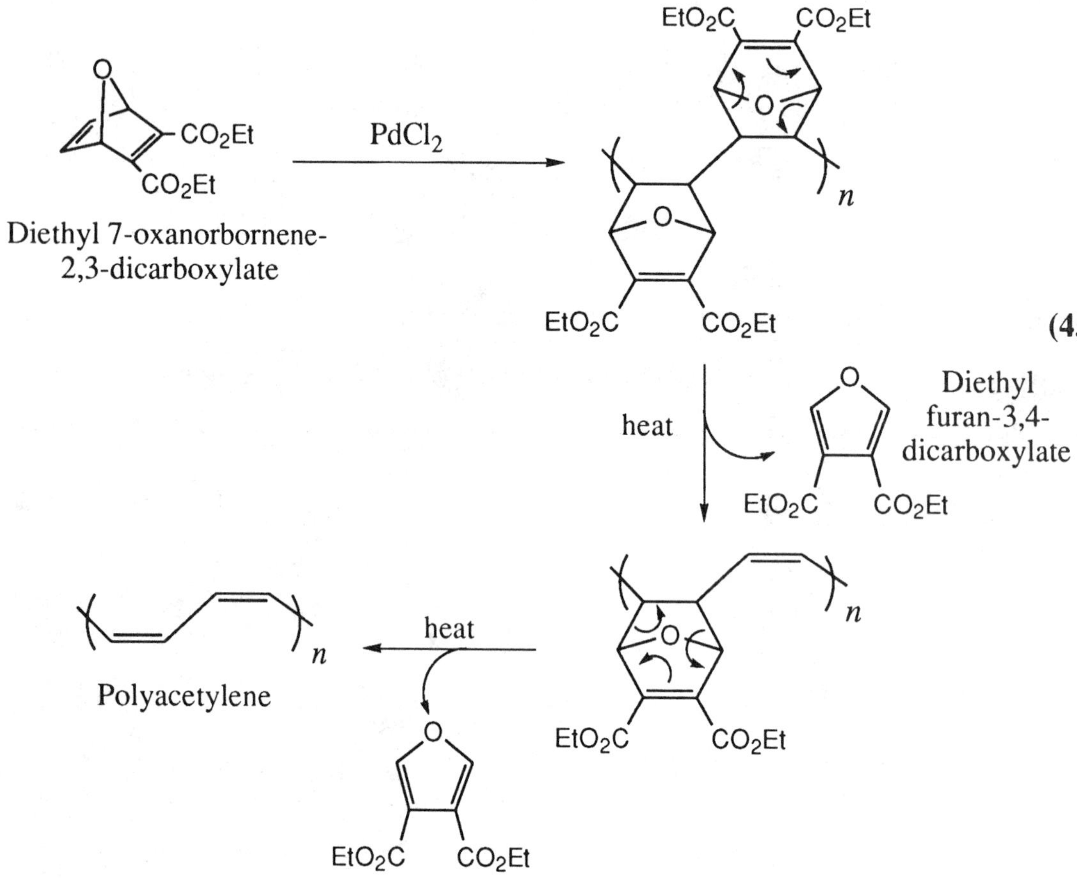
CO2Et
CO2Et
PdCl2
Diethyl 7-oxanorbornene-2,3-dicarboxylate
EtO2C
CO2Et
n
heat
Diethyl furan-3,4-dicarboxylate
Polyacetylene

(45)

Another important polymer in electro-optical applications is poly(phenylene vinylene) (PPV), which has alternating vinyl and phenyl groups.

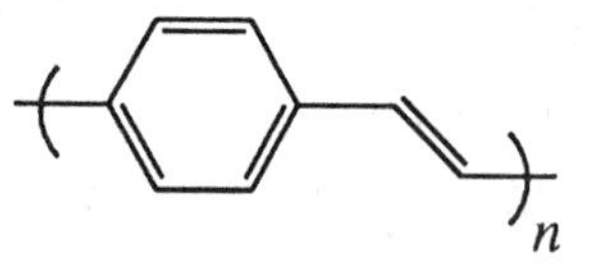

Poly(phenylene vinylene)

Precursor routes into this polymer have also been explored. Through the application of ROMP techniques, bicyclo[2.2.2]octadiene monomers substituted with acetate groups can be polymerized to form soluble, processable polymers. Heating these precursor polymers results in the elimination of two equivalents of acetic acid, which aromatizes the six-membered rings and completes the conjugation (equation 46).

Bicyclo[2.2.2]octadiene-
1,2-diacetate

Poly(phenylene vinylene)

(46)

Chemistry in Action

Recycling of Plastics

Polymers, in the form of plastics, are materials upon which our society is incredibly dependent. Durable and lightweight, plastic resins are probably the most versatile synthetic materials in existence; in fact, their current production in the United States exceeds that of steel. Plastics have recently come under criticism, however, for their role in the garbage crisis. They comprise 21% of the volume and 8% of the weight of solid wastes, most of which is derived from disposable packaging and wrappings. Of the 53 billion pounds of thermoplastic resins produced in 1993 in America, less than 2% was recycled.

Why aren't more plastics being recycled? The durability and chemical inertness of most resins make them ideally suited for reuse. The answer to this question has more to do with economics and consumer habits than with technological obstacles. Since curbside pickup and centralized drop-off stations for recyclables are just now becoming common, the amount of used material available for reprocessing has traditionally been small. This limitation, combined with the need for an additional sorting and separation step, rendered the use of recycled plastics in manufacturing expensive compared with virgin materials. Until recently, consumers perceived products made with "used" materials as being inferior to new ones, so the market for recycled products has not been large. However, the growth of the environmental movement over the last few years has resulted in a greater supply of used materials and a greater demand for recycled products. As manufacturers adapt to satisfy this new market, plastic recycling will eventually catch up to the recycling of other materials, such as glass and aluminum.

Six types of polymeric resins are commonly used for packaging applications. In 1988 many manufacturers began to use number codes, developed by the Society of the Plastics

Industry, on these products to facilitate recycling by consumers (Table 1). Because the plastics recycling industry still is not fully developed, only polyethylene terephthalate (PET) and high-density polyethylene (HDPE) are currently being recycled in large quantities, although outlets for the other resins are being developed (PET soft drink bottles are currently being recycled at a rate of 30%). Low-density polyethylene, which accounts for about 40% of plastic trash, has been slow in finding acceptance with recyclers. Facilities for the reprocessing of polyvinyl chloride (PVC), polypropylene, and polystyrene exist, but are still rare.

The process for the recycling of most plastics is simple, with separation of the desired resins from other contaminants the most labor-intensive step. A PET soft drink bottle with an HDPE base, for example, usually has a paper label, adhesive, and an aluminum cap that must be removed before the plastic can be reused. The recycling process begins with hand or machine sorting, after which the bottles are shredded into small chips. An air cyclone is then used to remove paper and other lightweight materials from the plastics. Any remaining labels and adhesives are eliminated with a detergent wash, and the PET and HDPE chips, which have different densities, are separated by flotation methods or with a hydrocyclone. Aluminum, the final contaminant, is removed electrostatically after the chips have been dried. The granulated plastics produced by this method are 99.9% free of contaminants and sell for about half the price of the unused materials. Unfortunately, plastics with similar densities cannot be separated with this technology, nor can resins composed of several polymers be broken down into pure components. However, recycled mixed resins can easily be molded into plastic lumber that is strong, durable, and graffiti-resistant. Several small companies now manufacture this lumber for use in park benches and picnic tables.

An alternative to this process, which uses only physical methods of purification, is **chemical recycling**. Eastman Kodak salvages large amounts of its PET film scrap using this technology. In a process known as *methanolysis*, methanol and a metal catalyst are added to the scrap. The methanol reacts with the polymer in a transesterification reaction, liberating ethylene glycol and dimethyl terephthalate.

PET   Ethylene   Dimethyl terephthalate
           glycol

These monomers are purified by distillation or recrystallization and used as feedstocks for the production of more PET film. Freeman Chemical Corporation uses a variation of this method, *glycolysis*, to produce short-chain polyalcohols which are then used in foam insulation or to make new polyester resins.

Table 1

Recycling Code	Polymer	Common Uses	Uses of Recycled Polymer
1 PET (or PETE)	Poly(ethylene terephthalate)	Soft drink bottles household chemical bottles	Soft drink and household chemical bottles
2 HDPE	High-density polyethylene	Milk and water jugs, grocery bags, bottles	Bottles, molded containers
3 V	Poly(vinyl chloride) (Vinyl or PVC)	Shampoo bottles pipes, shower curtains vinyl siding, wire insulation, floor tiles, credit cards	Plastic floor mats
4 LDPE	Low-density polyethylene	Shrink wrap, trash and grocery bags, sandwich bags, squeeze bottles	Trash and grocery bags
5 PP	Polypropylene	Plastic lids, clothing fibers, bottle caps, toys, diaper linings	Mixed resin component
6 PS	Polystyrene	Styrofoam cups, egg cartons, disposable utensils, packaging materials, appliances	Molded items such as cafeteria trays, rulers, frisbees, trash cans, videocassettes
7	All other plastics and mixed resins	Various	Plastic lumber, playground equipment, road reflectors

Summary

Polymers are an important class of materials that can be utilized in a wide range of applications. The properties of these materials depend heavily upon several factors, including the structure of the repeat unit, the chain architecture, the presence or absence of crystalline phases, the tacticity, inter-chain order and packing, and the materials' morphology. Polymers can be classified as either step-growth polymers or chain-growth polymers depending upon the type of synthetic route leading to them. **Step-growth polymerizations** involve the stepwise addition of difunctional monomers, with either A-B or A-A + B-B type combinations possible. Further structural variations such as crosslinks and branches can be introduced into the resulting polymer by the addition of multifunctional monomers to the reaction mixture. The formation of high-molecular weight polymers from step-growth processes requires the use of reactions that proceed with very high conversions. Important commercial polymers synthesized through step-growth processes include polyesters, polyamides, polyurethanes, and polyimides.

Chain-growth polymerization proceeds by the sequential addition of monomers to an active chain end. Important mechanisms for chain-growth polymerizations include free-radical, anionic, cationic, and transition metal-mediated polymerizations. In all cases, a chain reaction, which is comprised of initiation, propagation, and termination steps occurs. **Chain-transfer steps** that act to terminate one chain but simultaneously initiate the growth of another, and hence support the chain-reaction mechanism, may also be present. Chain polymerizations that proceed without chain-transfer or chain-termination steps are called **"living"** and are important for the formation of specialty polymers such as **diblock**, **triblock**, and **telechelic polymers**. Among the transition metal-mediated polymerizations, the **Ziegler-Natta polymerization** of ethylene and propylene are probably the most significant. These reactions proceed with high specificity to yield polymers that are stereoregular and highly linear (i.e., they lack significant branching). This regularity leads to highly crystalline polymers. When the chains are elongated and oriented through special processing procedures, a polymer with strength and stiffness greater than steel can be obtained. Another metal-mediated polymerization of some importance is the **ring-opening metathesis polymerization (ROMP)** of strained cyclic alkenes. ROMP is unusual in that all of the unsaturation present in the monomer is conserved in the resulting polymer. Hence, ROMP reactions are ideal for the formation of unsaturated or fully conjugated materials.

SUMMARY OF KEY REACTIONS

Step-Growth Polymerizations

Polyesters

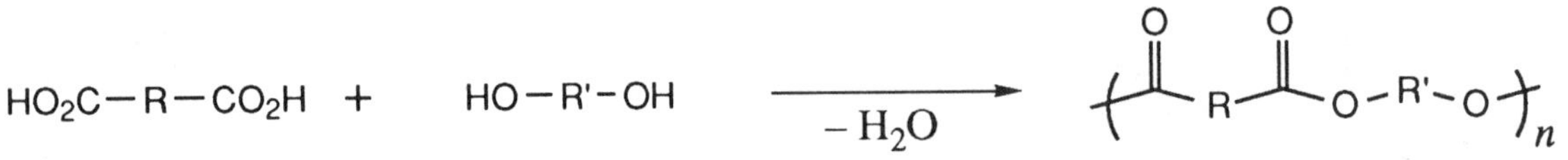

$$HO_2C-R-CO_2H \ + \ HO-R'-OH \ \xrightarrow[-H_2O]{} \ \left[\overset{O}{\overset{\|}{C}}-R-\overset{O}{\overset{\|}{C}}-O-R'-O \right]_n$$

Polyamides

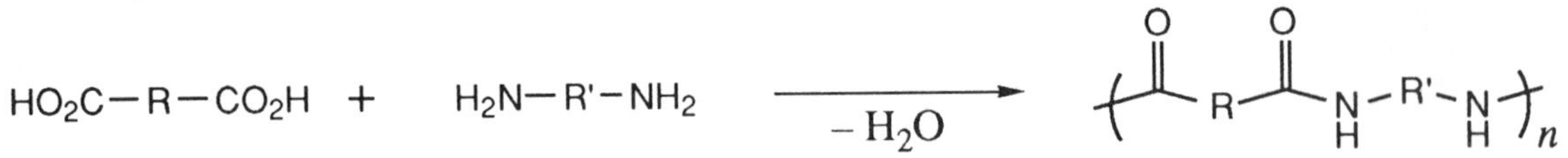

$$HO_2C-R-CO_2H \ + \ H_2N-R'-NH_2 \ \xrightarrow[-H_2O]{} \ \left[\overset{O}{\overset{\|}{C}}-R-\overset{O}{\overset{\|}{C}}-\underset{H}{N}-R'-\underset{H}{N} \right]_n$$

Polyurethanes

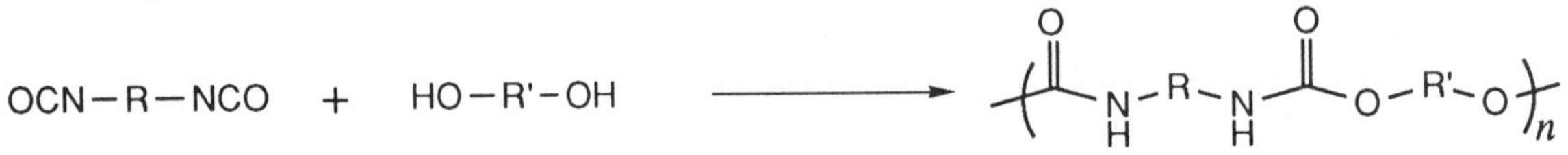

$$OCN-R-NCO \ + \ HO-R'-OH \ \longrightarrow \ \left[\overset{O}{\overset{\|}{C}}-\underset{H}{N}-R-\underset{H}{N}-\overset{O}{\overset{\|}{C}}-O-R'-O \right]_n$$

Epoxy-Resins

$$ \text{epoxide}-R-\text{epoxide} \ + \ H_2N-R'-NH_2 \ \longrightarrow \ \left(\underset{H}{N}-CH_2-\underset{OH}{CH}-R-\underset{OH}{CH}-CH_2-\underset{H}{N}-R' \right)_n$$

Polycarbonates

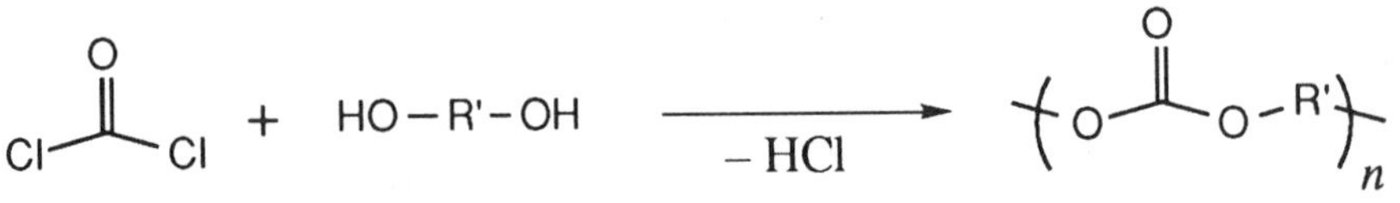

$$ Cl-\overset{O}{\overset{\|}{C}}-Cl \ + \ HO-R'-OH \ \xrightarrow[-HCl]{} \ \left[O-\overset{O}{\overset{\|}{C}}-O-R' \right]_n$$

Polyimides

$$ \text{dianhydride} \ + \ H_2N-R'-NH_2 \ \xrightarrow[-H_2O]{} \ \left(\text{imide}-R-\text{imide}-N-R' \right)_n$$

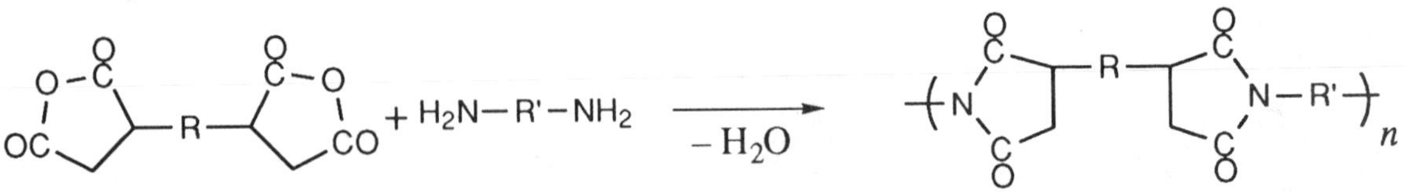

Polyethers

$$ n/2\,O_2 \ + \ \text{(2,6-disubstituted phenol)} \ \xrightarrow[-H_2O]{Cu^+,\ amine} \ \left[\text{(2,6-disubstituted phenylene)}-O \right]_n$$

70

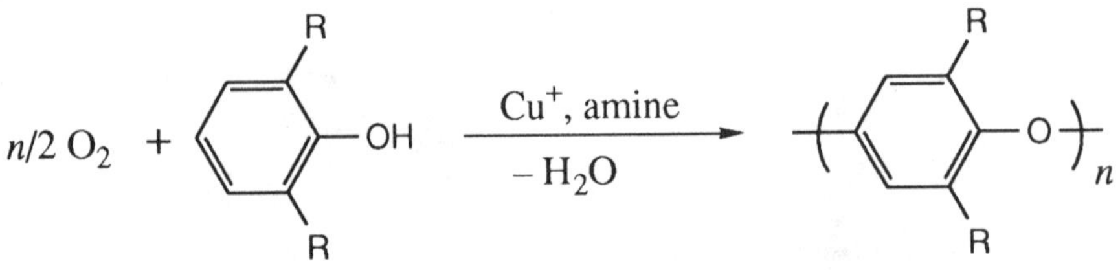

Polysulfones

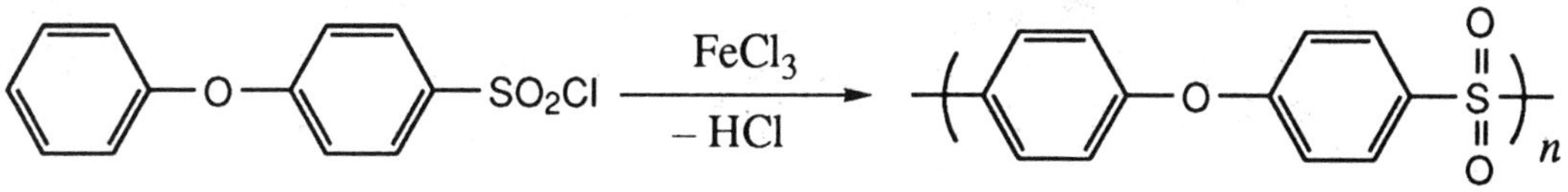

Chain-Growth Polymerizations

Anionic

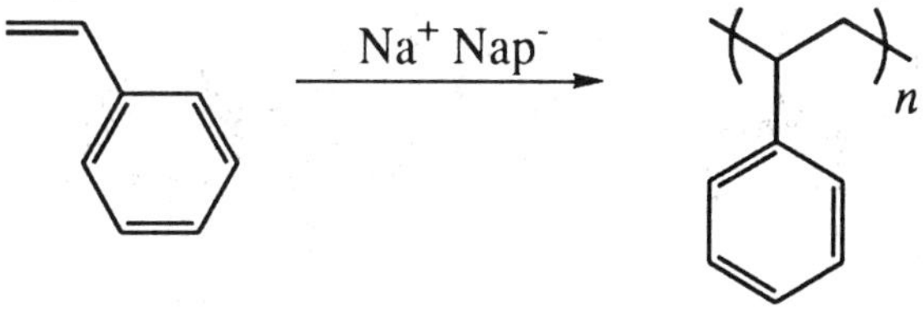

Cationic

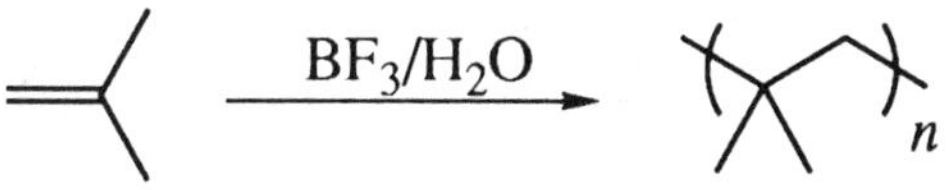

Free-Radical

Ziegler-Natta

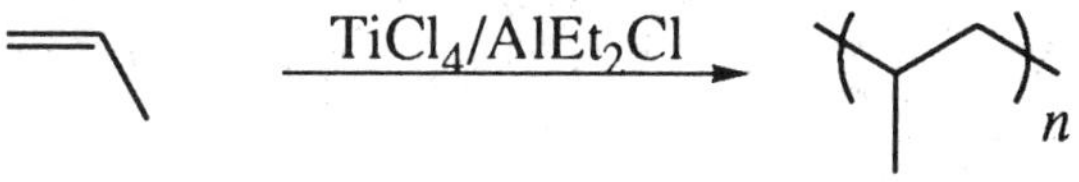

Ring-Opening Metathesis Polymerizations

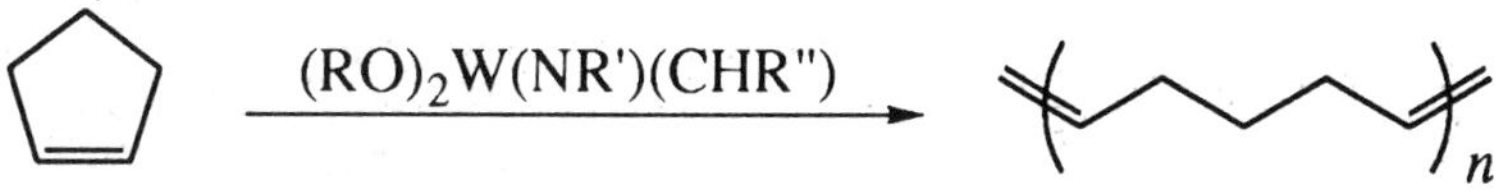

71

ADDITIONAL PROBLEMS

8.   Name the following polymers:

a.

b.

c.

d.

e.

f.

g.

h.

i.

j.

k.

l.

9.	Classify each polymer in Problem 8 as a condensation or addition polymer. Draw the structure(s) of the monomer(s) used to make each polymer.

10.	Given that a sample of poly(ethyleneterephthalate) (PET) has a number average molecular weight of 12 000, calculate its number average degree of polymerization, "n".

11.	Using the following data, calculate the number average, and weight average molecular weights (M_n and M_w) of this polymer sample. Using these results, calculate the polydispersity index.

Fraction	Mole Fraction	Molecular Weight
1	0.02	172,000
2	0.05	244,000
3	0.09	381,000
4	0.10	523,000
5	0.20	615,000
6	0.22	779,000
7	0.10	852,000
8	0.08	900,000
9	0.06	1,100,000
10	0.05	1,210,000
11	0.03	1,331,000

12.	How would you determine experimentally if a particular polymerization process is propagating by a step- or chain-growth mechanism?

13.	How would you prepare a block copolymer having blocks with the following structures?

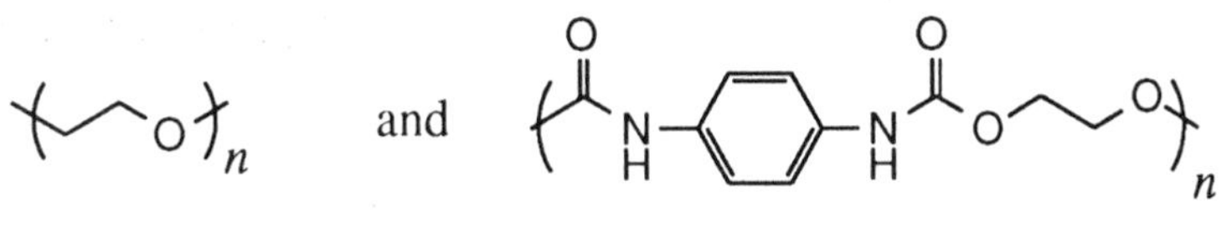

and

14. Draw the structure of the polymers obtained from the following reactions.

a.

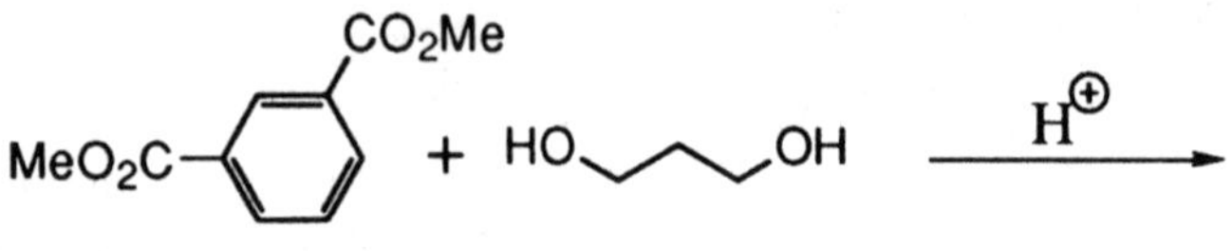

b.

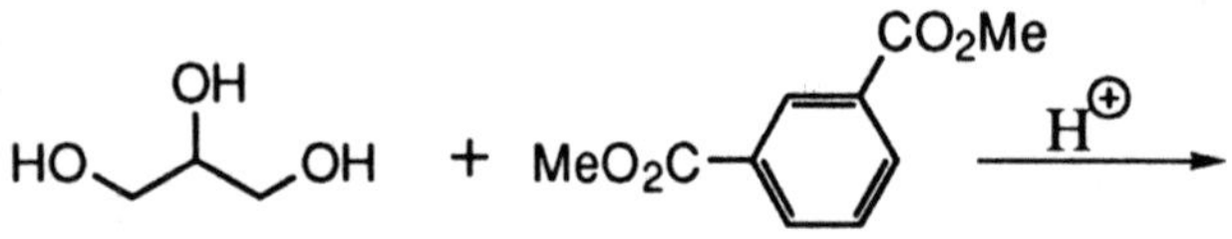

c.

AIBN
70 °C

d.

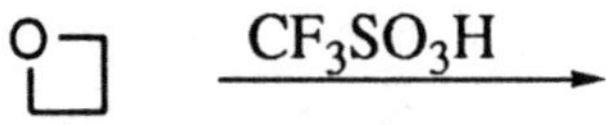

CF$_3$SO$_3$H

e.

KOH

f.

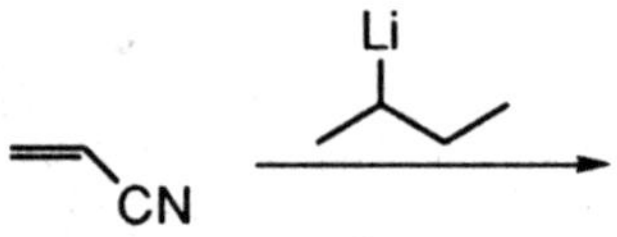

15. Chain transfer to monomer in cationic isobutylene polymerizations yields polymer chains with unsaturated endgroups.

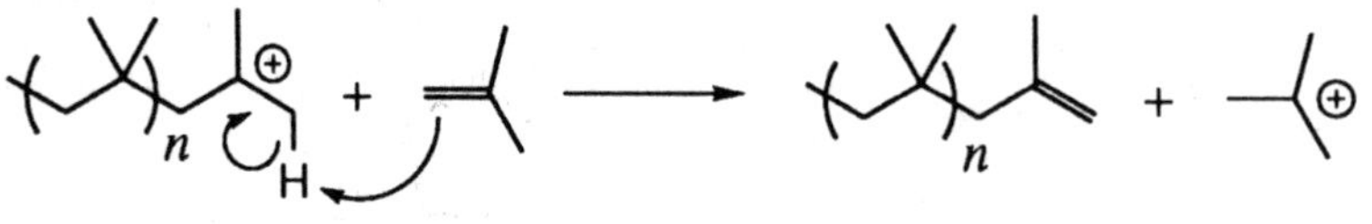

Given that 0.56 g Br$_2$ reacts completely with 5.0 g of polyisobutylene, calculate the average molecular weight of the polymer (M_n) and its average degree of polymerization, "n".

16.   Draw the structures of the polymers that would result from the ring-opening metathesis polymerization (ROMP) of the following monomers:

a.

b.

c.

d.

17.   A "double isomerization" polymerization of 2-pyrrolidino-2-oxazoline using benzyl chloride as the initiator has recently been reported to yield polymer **I**. In contrast, polymer **II** is obtained if trifluoromethyl methanesulfonate is used as the initiator.

$$CH_3SO_3CF_3$$ (II)

$$PhCH_2Cl$$ (I)

Outline the cationic polymerization mechanism of 2-pyrrolidino-2-oxazoline. Use this scheme to explain why one route occurs if trifluoromethyl methanesulfonate is used and the other occurs if PhCH$_2$Cl is used.

18.   Arrange the following monomers in decreasing order of reactivity in cationic polymerizations. Briefly explain your prediction.

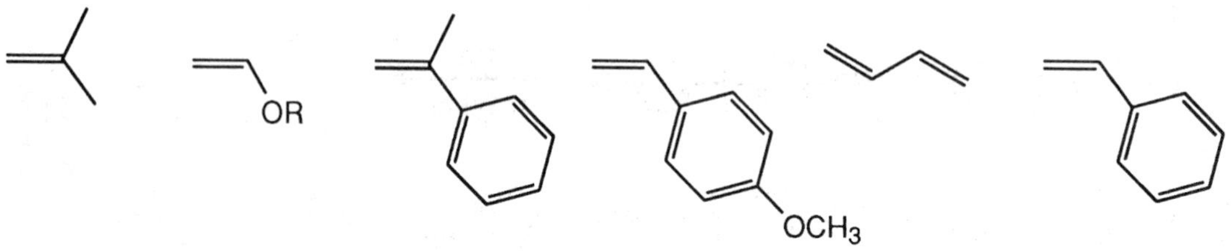

19.   Discuss the differences between step- and chain-growth polymerization mechanisms.

20.   Although most reaction rates increase with increasing temperature, the rates of many anionic polymerizations decrease with increasing temperature. (This is sometimes referred to as a "negative activation energy.") How might you account for this unusual behavior.

21.   a) Give the product polymer resulting from each of the following sets of reactions (you need only to draw the structure once). Clearly label the equations with "N. R." if no reaction will occur. b) In each of the pairs of reactions shown, predict which reaction should go faster. Briefly explain your answer.

i.

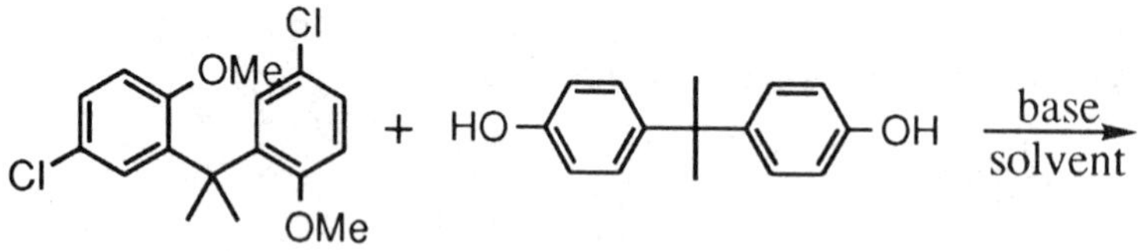

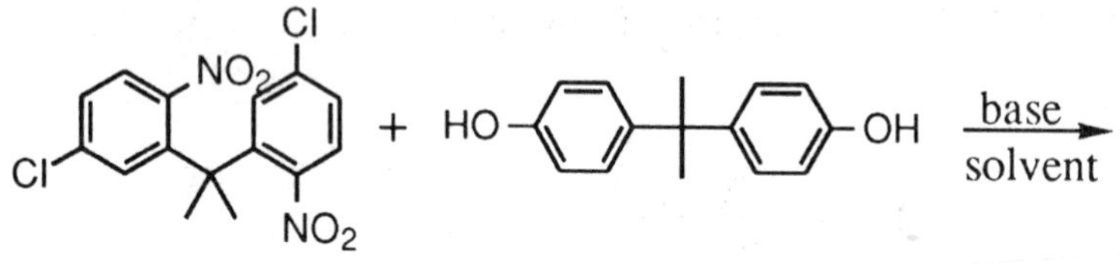

ii.

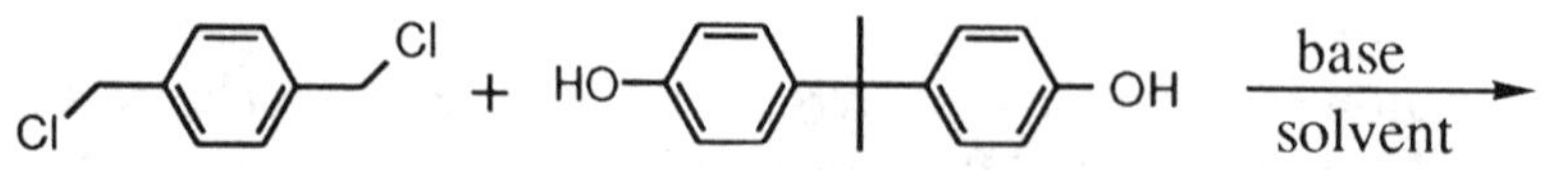

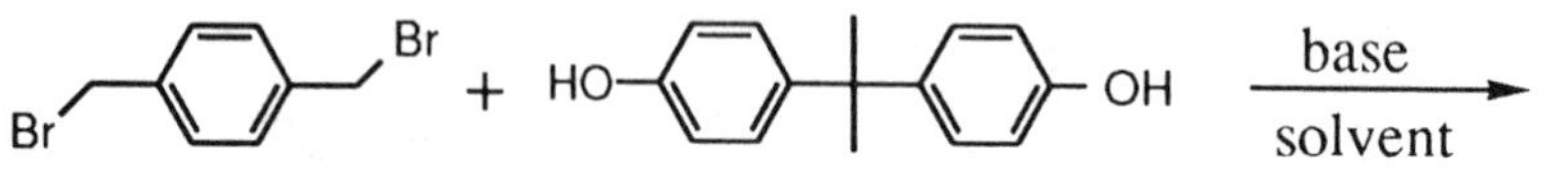

Problem 21 (Continued).

iii.

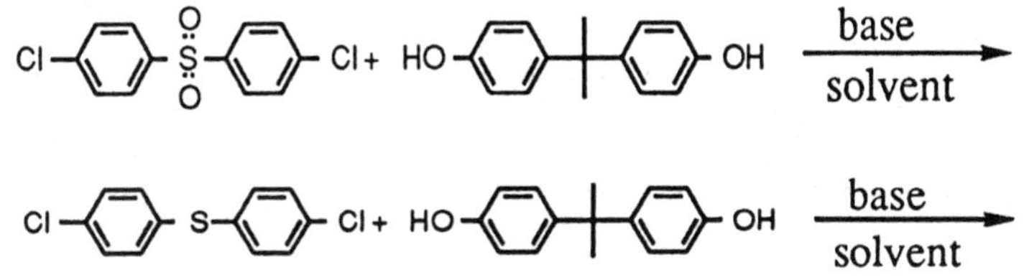

iv.

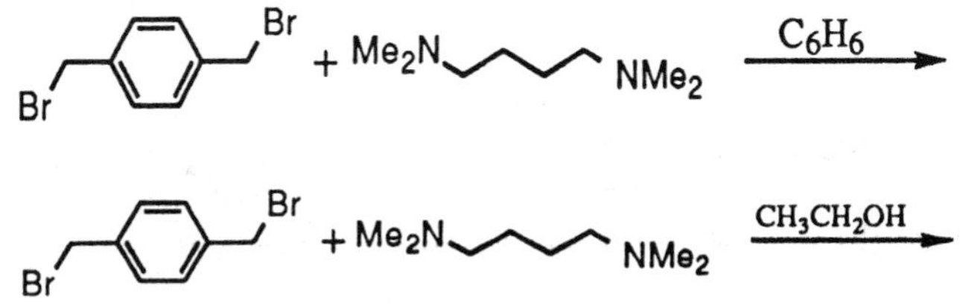

22. Most step-polymerizations involve equilibrium reactions, which become important for closed systems. Consider an acid-catalyzed polyesterification in a closed system with excess HCl added:

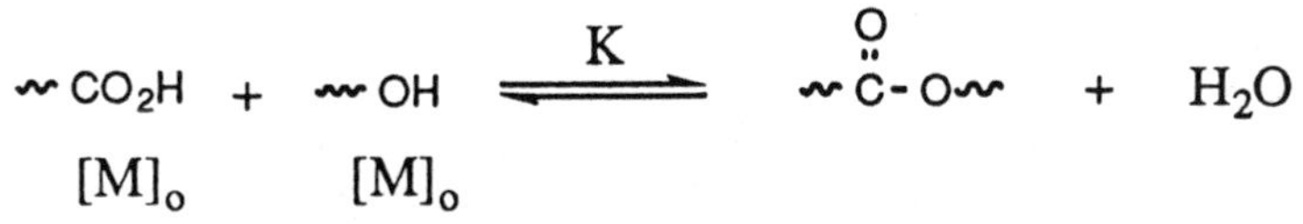

Assuming equal initial concentrations of carboxylic acid and alcohol functional groups, and using the relationship $X_n = 1/(1 - p)$, where p is the extent of reaction (fraction of the monomers reacted), show that:

$$X_n = \frac{K - 1}{K^{1/2} - 1}$$

23.   Propose a detailed mechanism to account for the following polymerization.

24.   Write a resonable mechanism for the ring-opening polymerization of caprolactam under acidic conditions.

25.   Explain why isobutylene (2-methylpropene) can only be polymerized under cationic conditions. What happens under free-radical or anionic conditions?